BUILDING THE DIGITAL WORKFORCE

STRATEGIES FOR AGENTIC AI SUCCESS

Acknowledgements

Edited by: Sinchana Mistry

Research support:

Anthropic Claude Sonnet 4, Opus 4 and Opus 4.1

Perplexity

OpenAI ChatGPT 4o

www.yourdigitalworkforce.com

Images:
Napkin AI

Canva

Forward

By Sridhar Vembu, Co-founder and CTO, Zoho Corp.

As I write this in September 2025, the most important decision facing businesses everywhere is how best to embrace, adopt and adapt to AI.

AI agents are top of mind for every decision maker. Navigating through a world where years of development now unfold in weeks is uniquely difficult.

This book very capably lays out the landscape and offers valuable guide posts.

The important point is to conduct small, focused experiments and learn.

Our own experience so far at Zoho, in this rapidly evolving landscape, is that AI agents are helpful if we never forget the "Prime Directive," that humans have to accept responsibility for their actions and humans have to be in charge.

That message is often lost in vendor marketing hype but it needs to be stressed.

CONTENTS

INTRODUCTION:
THE EMERGING DIGITAL WORKFORCE

At 3:47 AM on a Tuesday, Maya detected an unusual pattern in user complaints about a software feature. While the support team was offline, she analyzed thousands of tickets, identified the root cause in a recent code deployment, and coordinated an automated fix. She scheduled a patch deployment and pro-actively sent personalized updates to affected customers. By morning, what could have been a company crisis was resolved and documented.

Maya isn't human; she's an AI agent operating with the auton-omy and judgment once thought to be uniquely human. She represents a workforce revolution that's already reshaping how the world's most innovative companies operate.

The Digital Workforce Revolution

We stand at the brink of a disruptive shift in how work gets done. Throughout history, the nature of work has evolved through distinct phases—from manual labor in agrarian societ-ies, to mechanized production during the Industrial Revolution, to knowledge work in the Information Age.

Today, we face what may be the most significant shift yet: the rise of the digital workforce.

The digital workforce represents a fundamental departure from traditional workforce models, not simply another wave of automation. It is a shift in which artificial intelligence (AI) transitions from being merely a tool that requires constant human guidance to becoming a collaborative partner that demonstrates autonomous agency.

Organizations that once defined their workforce strictly as human resources are now expanding their definition to include *AI agents* that operate with increasing levels of independence. They are reshaping our understanding of what constitutes a "worker" in the modern economy.

Why Agentic AI?

The concept of AI as an "agent" is an evolutionary step in its development. Traditional AI systems have operated as passive instruments that augment human capabilities.

Agentic AI, by contrast, possesses a degree of autonomy that allows it to act on behalf of humans or organizations toward defined objectives.

What makes an AI system "agentic" is its capacity to:

- **Perceive** its environment through various forms of data input and process this information to understand context
- **Decide** by planning sequences of actions to achieve specified goals
- **Act** by executing these actions with limited or no human intervention
- **Learn** from outcomes to improve future performance

Cycle of Agentic AI

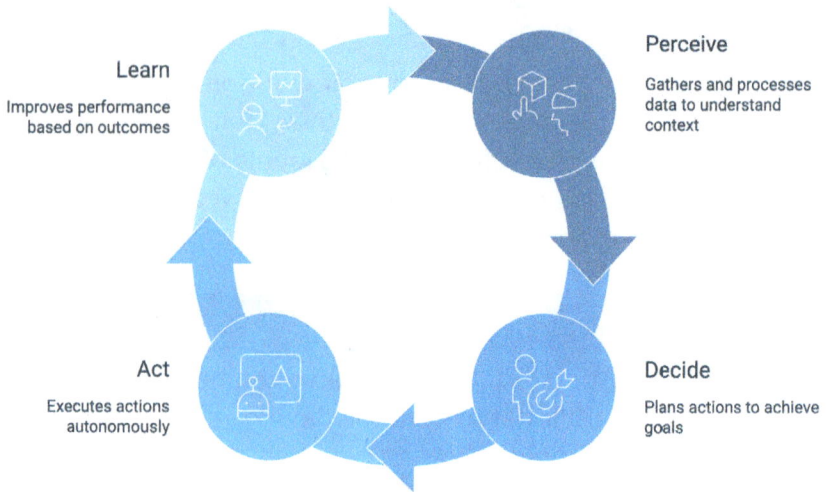

Learn
Improves performance based on outcomes

Perceive
Gathers and processes data to understand context

Act
Executes actions autonomously

Decide
Plans actions to achieve goals

This autonomy (*or agency*) transforms the relationship between humans and AI from one of constant supervision to one of delegation and collaboration. Agentic AI can be assigned objectives and trusted to determine appropriate paths to achieve them, much as we would trust a human colleague.

The business implications of this shift are profound. Organizations are now able to deploy digital workers to handle increasingly complex workflows that require judgment, adaptability, and contextual understanding. Whether managing customer interactions, orchestrating supply chains, analyzing market trends, or optimizing operations, agentic AI is redefining what is possible without direct human involvement.

The Promise and Challenge of AI

The digital workforce offers unprecedented opportunities for organizations across every industry sector and business function. The potential benefits include:

1. **Scalability beyond human constraints:** Digital workers operate 24/7 without fatigue, scale horizontally to meet

demand surges, and process information volumes that would overwhelm human teams.

2. **Operational consistency:** Once properly configured, AI agents deliver consistent quality and adherence to protocols without the variability inherent in human performance.

3. **Accelerated innovation cycles:** Digital workers free human talent to focus on creative problem-solving, strategic thinking, and innovation.

4. **Economic transformation:** The cost economics of digital labor promises to dramatically reduce operational expenses while simultaneously improving service quality and availability.

5. **Enhanced human capabilities:** The most powerful applications pair digital and human workers in collaborative relationships that amplify the unique strengths of each.

Yet alongside these opportunities lie significant challenges that organizations must navigate:

1. **Technical complexity:** Implementing effective agentic AI systems requires sophisticated technical infrastructure, data quality management, and integration capabilities that many organizations currently lack.

2. **Ethical considerations:** As AI assumes greater autonomy, questions of responsibility, accountability, and alignment with human values become urgent.

3. **Workforce transformation:** The integration of digital workers requires reimagining organizational structures, job roles, and management approaches.

4. **Security and resilience:** Systems with greater autonomy introduce new vulnerabilities that must be proactively addressed through robust security frameworks.

5. **Governance frameworks:** Organizations must develop new governance models appropriate for overseeing entities that blend characteristics of both tools and independent actors.

The urgency for executives to embrace this shift cannot be overstated. Early adopters of well-implemented digital workforce strategies will gain decisive competitive advantages in the coming decades.

Navigating This Book

This book aims to provide a comprehensive framework for organizations seeking to build and deploy effective digital workforces. In the chapters that follow, we will explore:

- The technological foundations of agentic AI and how they enable autonomous action
- Strategic approaches to identifying high-value applications for digital workers
- Implementation methodologies that maximize success while minimizing disruption
- Management principles for supervising and collaborating with digital team members
- Ethical frameworks for ensuring the responsible deployment of autonomous systems
- Future trajectories of the digital workforce evolution and how to prepare for them

The transformation to hybrid (digital-human) workforces is not merely a technological challenge but an organizational and cultural shift. It requires rethinking basic assumptions about how work gets done, how value is created, and how human and digital capabilities can be most effectively combined.

The organizations that will thrive in this new era will be those that approach the digital workforce not simply as a cost-cutting measure but as a strategic imperative—one that demands thoughtful leadership, cultural adaptation, and a willingness to reimagine core business processes. This book provides the roadmap for that journey.

PART
ONE
FOUNDATIONS OF
THE DIGITAL WORKFORCE

CHAPTER 1:

UNDERSTANDING AGENTIC AI

TL;DR:

- Agentic AI is an evolution beyond traditional automation, characterized by goal-directed behavior, environmental interaction, and adaptation.

- The technical foundations of today's AI systems include foundation models, reasoning capabilities, tool use, and sophisticated context management.

- Current systems excel at information synthesis, content creation, process execution, and knowledge work assistance, though important limitations remain.

- The most effective implementations of agentic AI recognize both the capabilities and limitations of current systems, creating collaborative relationships between human and digital workers.

- Organizations that develop a nuanced understanding of agentic AI will capture its significant strategic value while avoiding implementation pitfalls.

From Automation to Agency: The Evolution of AI Capabilities

The journey toward agentic AI is one of the most significant technological progressions of our time. To fully appreciate where we stand today and the transformative potential that lies ahead, we must first understand the evolutionary path that brought us here.

This evolution has unfolded through distinct phases, each building upon the capabilities of its predecessors while introducing impactful new possibilities.

The Era of Rule-Based Automation

The earliest forms of workplace automation systems, emerging in the 1960s and 1970s, followed explicitly programmed rules and predetermined workflows.

Early banking software exemplified this approach, processing transactions with precision and consistency, but only within narrowly defined parameters. These systems were brittle; they could only handle scenarios their programmers had explicitly anticipated. Any deviation from expected inputs would result in errors or system failures rather than adaptation.

The limitations of rule-based systems became increasingly apparent as organizations sought to automate more complex processes. These systems lacked the flexibility to navigate real-world ambiguity, could not learn from experience, and required extensive human oversight to maintain and update.

The Rise of Machine Learning

The next major evolution came with the mainstream adoption of machine learning techniques in the 1990s and early 2000s. Unlike rule-based systems, machine learning models could identify patterns in data and make predictions based on statistical relationships. This was a large shift in how AI functioned, from purely deterministic programming to probabilistic learning.

Early applications included fraud detection systems identifying suspicious patterns in financial transactions, product recommendation engines based on user behavior, and quality control systems spotting manufacturing defects.

These applications demonstrated a new kind of flexibility, the ability to adapt and improve based on data rather than explicit programming.

While machine learning systems could find patterns humans might miss and improve their performance over time, they remained tools rather than agents. They lacked the ability to take autonomous action based on their insights, requiring humans to interpret outputs and implement decisions.

The Deep Learning Revolution

The 2010s saw the emergence of deep learning as a transformative approach to AI. Neural networks with multiple layers could recognize complex patterns in unstructured data, such as images, audio, and natural language, enabling systems to perceive and interpret the world in ways more closely resembling human cognition.

Computer vision systems achieved near-human accuracy in object identification. Natural language processing models could extract meaning from text and generate coherent language. Speech recognition reached levels of performance that made voice interfaces practical for everyday use.

These capabilities represented a crucial stepping stone toward agency, the ability to perceive and make sense of complex, unstructured environments. Yet even these powerful systems remained reactive, processing inputs and generating outputs without independent goal-directed behavior.

Foundation Models

The late 2010s and early 2020s witnessed another paradigm shift with the development of what we now call foundation models: large-scale AI systems trained on massive datasets and adaptable across a wide range of tasks.

These include large language models (LLMs) like OpenAI's ChatGPT and Anthropic's Claude, multimodal models that process both text and images, and specialized systems for scientific applications.

Foundation models introduced capabilities crucial for agentic systems:

- **General-purpose reasoning** to tackle novel problems beyond pattern recognition
- **Contextual understanding** to maintain coherence over extended interactions
- **Instruction following** to respond appropriately to natural language directives
- **Knowledge integration** to apply broad world knowledge acquired during pre-training

These capabilities dramatically reduced the gap between human intention and machine action. For the first time, non-technical users could direct sophisticated AI systems through natural language alone, opening new possibilities for human-machine collaboration.

The Current Frontier: From Tools to Agents

Today, we enter a new phase in AI: the transition from tools to agents. This shift is characterized by the integration of key capabilities:

- **Goal-directed behavior** to pursue objectives over extended periods

- **Environmental interaction** to take actions that affect the world beyond generating text or images
- **Tool use** to employ specialized software, services, and knowledge bases
- **Planning and adaptation** to develop and revise sequences of actions as circumstances change

The difference between a sophisticated tool and an agent may appear subtle, but it is a profound shift in how organizations deploy AI.

Unlike traditional tools that require constant direction, agents pursue goals autonomously, transforming the human-AI relationship from one of supervision to one of delegation and collaboration.

Defining Agentic AI: Key Characteristics and Capabilities

What precisely makes an AI system "agentic" rather than merely automated? The distinction lies in a set of core capabilities that enable such systems to operate with autonomy, adaptability, and purpose.

Perception: Understanding the World

Unlike earlier AI systems, which relied on clean, structured inputs, agentic systems require sophisticated perception capabilities to interpret messy, real-world information across multiple modalities.

They comprehend natural language in documents and messages, perceive and interpret visual content like images, charts, and videos, and analyze structured data from spreadsheets and databases.

Advanced agentic systems integrate information across these modalities, enabling them to understand complex situations that no single data source could fully capture.

Planning: Charting Paths to Goals

Perhaps the most distinctive characteristic of agentic AI is its ability to plan: develop sequences of actions to achieve a desired objective.

These systems break down complex objectives into manageable subtasks, determine the appropriate sequence of actions, allocate time and computational resources effectively, and assess potential risks to create contingency plans.

Unlike traditional systems that follow predetermined workflows, agentic AI plans dynamically based on current circumstances, available resources, and specified objectives.

Execution: Taking Effective Action

Agentic systems distinguish themselves by executing plans and taking concrete actions in their environments.

They manipulate tools using specialized software, APIs, and databases, generate content, communicate with humans or other AI systems, and navigate digital environments to accomplish tasks.

What makes these execution capabilities truly agentic is that they are directed by goals. Actions are not taken randomly or in isolation but as part of coherent strategies to achieve specific outcomes.

Learning and Adapting: Improving Through Experience

The final critical characteristic of agentic AI is its capacity to learn and adapt over time. Through repeated execution, it refines its performance, enhances strategies based on outcomes, accumulates relevant knowledge, and becomes better at interpreting user preferences.

This continuous improvement means agentic systems grow more valuable over time, unlike traditional software that typically degrades gradually until the next update.

Key Attributes of AI Agents

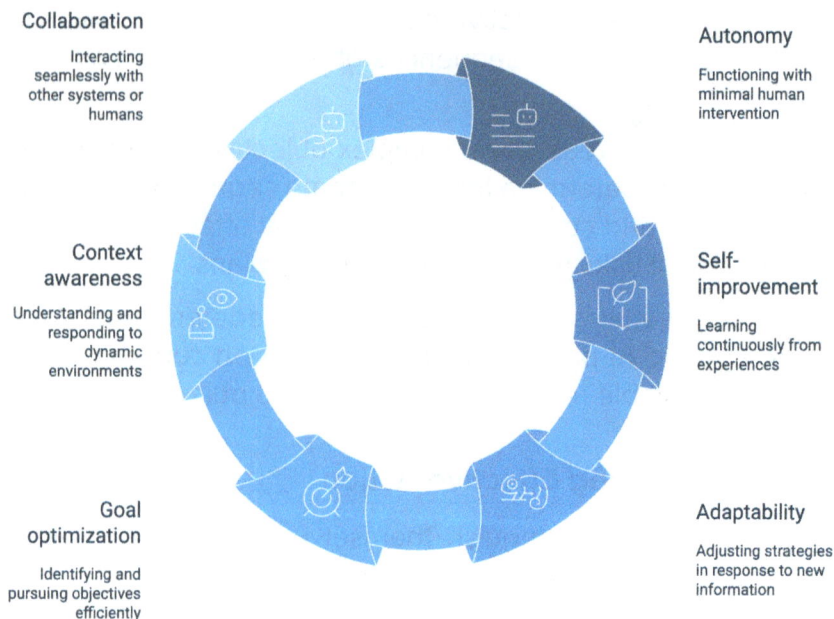

Collaboration
Interacting seamlessly with other systems or humans

Autonomy
Functioning with minimal human intervention

Context awareness
Understanding and responding to dynamic environments

Self-improvement
Learning continuously from experiences

Goal optimization
Identifying and pursuing objectives efficiently

Adaptability
Adjusting strategies in response to new information

Technical Foundations: The Building Blocks of Today's Agentic Systems

Understanding the technical foundations of agentic AI systems helps organizations make informed decisions about implementation strategies, priority use cases, capability needs, and where it may fall short.

While agentic AI continues to advance rapidly, several key components have emerged as essential building blocks.

Foundation Models: The Cognitive Core

At the heart of contemporary agentic AI systems lie foundation models, large-scale neural networks trained on massive datasets that provide core cognitive capabilities. These models have several crucial properties:

- **Generality**: Operating across diverse functions rather than being restricted to narrow tasks
- **Transferability**: Applying knowledge acquired in one context to new situations
- **Composability**: Combining with other components to create more capable systems
- **Scalability**: Improving capabilities with increases in model size, training data, and computational resources

Early foundation models focused on language processing, but today's specialized and multimodal systems handle images, audio, structured data, behavioral data, and other information types.

The availability of foundation models has accelerated the development of agentic AI by providing pre-trained cognitive capabilities that previously would have required years of specialized development. This has democratized access to sophisticated AI, allowing organizations without extensive machine learning expertise to deploy powerful systems.

Reasoning Capabilities: Beyond Pattern Recognition

While foundation models excel at pattern recognition and knowledge retrieval, agentic AI systems require more sophisticated reasoning capabilities.

Techniques such as chain-of-thought prompting and tree search allow models to explore possible solutions and articulate intermediate steps before committing to a course of action. External memory systems help maintain structured knowledge during problem-solving, while self-critique mechanisms allow agents to verify and refine their outputs.

These capabilities enable agentic systems to solve complex problems that require multi-step thinking, constraint handling, and logical inference.

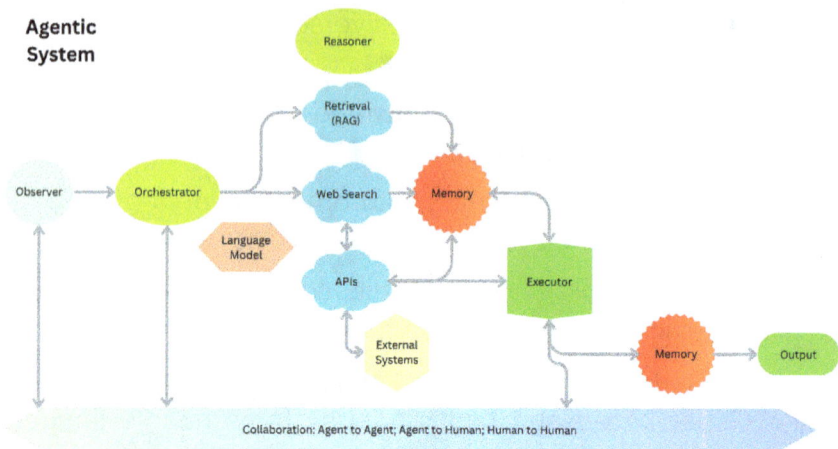

Agentic System

Observer → Orchestrator → Retrieval (RAG) / Web Search → Reasoner / Memory

Language Model

APIs → External Systems

Executor → Memory → Output

Collaboration: Agent to Agent; Agent to Human; Human to Human

Tool Use: Extending Capabilities Through Integration

A defining characteristic of advanced agentic systems is their ability to use external tools, specialized software, APIs, databases, and other resources.

External tool use is enabled through the use of standard protocols like Model Context Protocol (MCP). MCP is an open standard that allows AI assistants to securely connect to external data sources and tools through a standardized interface. This capability enables AI models to access real-time information from databases, APIs, and applications while maintaining security and consistent data formatting. This allows them to complement their general cognitive capabilities with specialized functions.

Agents can access real-time information that was unavailable during training, perform domain-specific calculations, interface with enterprise software, and take action across digital environments. A well-designed agent seamlessly integrates dozens or even hundreds of tools, knowing when and how to employ each based on the task at hand.

Context Management:
Maintaining Coherence Over Time

For AI systems to function as reliable agents rather than merely responsive tools, they must maintain coherent context over extended interactions. This capability has several dimensions.

It involves tracking conversation history, monitoring task progress, remembering user preferences, and remaining aware of evolving environmental constraints.

Advanced context management enables agentic systems to participate in extended collaborations, maintain consistency across interactions, and adapt their behavior based on accumulated experience with users and tasks.

Current State of the Art:
What's Possible Today vs. Common Misconceptions

As with any rapidly evolving technology, agentic AI is surrounded by both excessive hype and unwarranted skepticism. Organizations developing implementation strategies must understand what these systems can reliably accomplish today versus what remains difficult.

Current Capabilities: What Today's Systems Do Well

Today's agentic AI systems excel at several categories of tasks that were beyond the reach of automation just a few years ago:

- **Information synthesis**: Gathering, analyzing, and summarizing information from multiple sources
- **Content creation**: Generating high-quality text, images, and code based on specifications
- **Process execution**: Following complex workflows that involve multiple systems and decision points
- **Knowledge work assistance**: Supporting professional knowledge workers in industries like law, finance, and medicine

- **Customer interaction**: Engaging with customers to resolve issues, answer questions, and provide recommendations
- **Collaboration**: Interacting and collaborating with other agents and humans

These capabilities are already creating significant value in organizations that have deployed agentic systems effectively. The key is identifying appropriate use cases that match current capabilities rather than expecting these systems to solve every problem.

Common Limitations: Where Challenges Remain

Equally important is understanding the limitations of current systems to set realistic expectations and avoid implementation failures:

1. **Reliability at scale**: Decreasing reliability with increasing task complexity
2. **Common sense reasoning**: Understanding implicit physical and social constraints
3. **True creativity**: Generating truly innovative and original ideas
4. **Complex decision-making**: Making judgments involving multiple stakeholders, ethics, and long-term consequences
5. **Physical world interaction**: Operating directly in physical environments without human intermediation (though agents can be integrated with robots to perform such tasks)

These limitations don't diminish the value of current systems but do suggest appropriate boundaries for their application. The most successful implementations pair agentic AI with human collaborators who provide the judgment, creativity, and responsibility that remain beyond algorithmic capabilities.

Dispelling Common Misconceptions

Several persistent misconceptions about agentic AI lead organizations astray in their implementation strategies:

Misconception 1: Agentic AI systems either have human-level general intelligence or are worthless.
Reality: There is a huge and valuable middle ground of specialized capabilities that create enormous business value without approaching general intelligence.

Misconception 2: Implementing agentic AI requires rebuilding all existing systems.
Reality: The most successful approaches integrate agentic systems with existing infrastructure rather than replacing it entirely.

Misconception 3: Agentic systems are fully autonomous and require no human involvement.
Reality: Thoughtful human-AI collaboration remains essential for oversight, quality, and safety.

Misconception 4: Once deployed, agentic systems will automatically optimize themselves.
Reality: While these systems learn from experience, they require ongoing monitoring, evaluation, and refinement to maintain and improve performance.

Misconception 5: For an AI agent to be "agentic," it must be performing all the capabilities it possesses (or put another way, it's not agentic if it is being used to perform tasks that are not autonomous with full agency)
Reality: AI agents have the capabilities to perform a variety of use cases. Limited use doesn't negate their agentic nature.

Understanding these nuances helps organizations develop realistic implementation strategies that maximize the value of current capabilities while preparing for future advancements.

Near-Term Development Trajectory

Based on current research trends and technological progress, we can anticipate several important developments in agentic AI capabilities over the next one to three years:

- **Improved reliability**: Maintaining performance across a wider range of tasks and situations

- **Enhanced reasoning**: Solving problems more effectively, especially in structured contexts

- **Greater personalization**: Adapting more closely to individual user preferences and organizational settings

- **Expanded tool use**: Integrating seamlessly with specialized software and information sources

- **Improved collaboration**: Coordinating more naturally and effectively between human-AI team members and between AI agents

These advancements will not change the nature of agentic AI but will make existing capabilities more reliable, accessible, and applicable to a wider range of business contexts.

Conclusion: The Strategic Imperative

Understanding agentic AI, its evolution, defining characteristics, technical foundations, and current capabilities, provides the essential groundwork for strategic implementation.

This understanding allows organizations to:

1. **Identify appropriate use cases** that align with current capabilities
2. **Set realistic expectations** about what these systems can accomplish
3. **Develop effective integration approaches** that combine AI and human strengths
4. **Prepare for future developments** without being misled by hype or unfounded concerns

The transition from viewing AI as a tool to recognizing its potential for agency marks a fundamental change in how organizations approach automation and augmentation. Organizations can now begin to delegate entire workflows and processes to digital colleagues capable of navigating complexity with decreasing levels of human supervision.

CHAPTER 2:
THE ECONOMICS OF DIGITAL LABOR

TL;DR:

- Digital workers present a different economic profile from traditional labor with high fixed costs but minimal marginal costs, creating powerful scale economies.

- Their value extends far beyond labor replacement to include quality improvements, capacity expansion, and entirely new capabilities.

- Digital and human workers exhibit both substitution and complementarity effects, with the greatest value typically coming from enhanced human capabilities rather than simple replacement.

- The economic impact varies dramatically across sectors and job categories based on task characteristics, emotional intelligence requirements, and complexity.

- Digital workers enable business model innovations, including continuous service availability, new pricing approaches, and entirely new product categories.

Cost Economics of Digital Workers: Analyzing the Financial Implications

The economics of digital labor is one of the most compelling aspects of the agentic AI revolution. Unlike traditional labor economics, digital workers follow a radically different financial model that organizations must understand to develop sound investment strategies.

Fixed vs. Variable Costs in Digital Workforce Deployment

Traditional human workforces operate on predominantly variable cost structures. Each additional worker requires a proportional increase in compensation, benefits, workspace, equipment, and management overhead.

This offers organizations the financial flexibility to scale their workforce up or down, but it also results in labor costs increasing linearly with capacity.

Digital workers present a dramatically different cost profile:

High fixed costs: The initial development, customization, and deployment of agentic AI systems involve substantial investment.

This includes licensing or fine-tuning foundation models, integrating them with existing systems, designing and optimizing processes, preparing data, training and change management, and developing governance infrastructure.

Low marginal costs: Once deployed, the cost of scaling digital workers is minimal compared to human labor. Each additional "instance" of a digital worker requires incremental computational resources, minimal additional storage, and negligible additional maintenance.

This creates a cost curve that looks different from traditional labor economics—high initial investment followed by rapidly declining per-unit costs as scale increases.

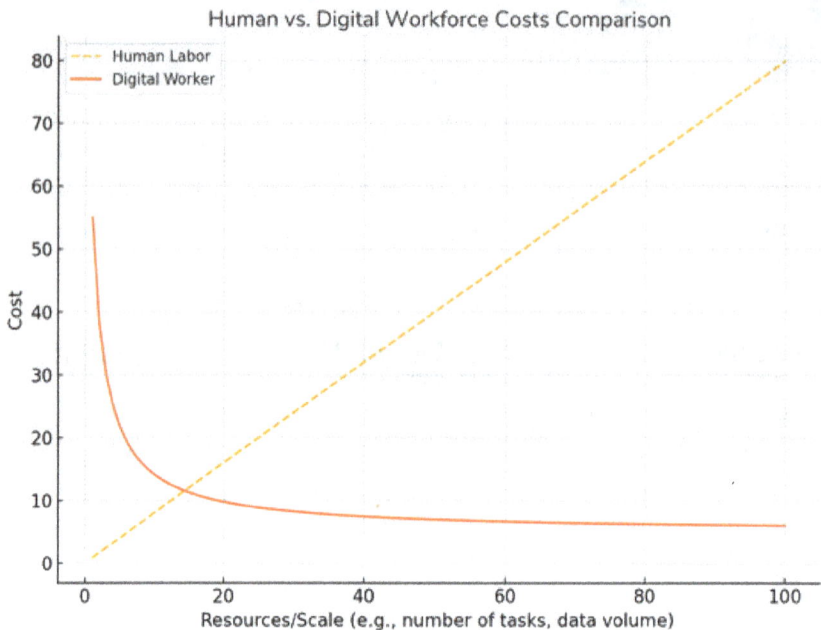

Human vs. Digital Workforce Costs Comparison

This shift requires organizations to rethink workforce planning by identifying processes with sufficient volume to justify initial investments and offer significant economies of scale.

Economies of Scale and Network Effects

Digital workforces exhibit powerful economies of scale and network effects that compound as adoption grows.

Economies of scale play a significant role as organizations expand their digital workforces. Fixed development costs are spread across a growing number of use cases, making each subsequent deployment more cost-effective.

Organizations also benefit from common components and infrastructure, institutional knowledge that accelerates new deployments, favorable negotiating terms with technology providers, and process optimization based on accumulated data.

Network effects arise because, unlike traditional workers, whose knowledge and outputs often remain siloed, digital workers immediately transfer capabilities developed for one process to others within an organization.

Insights gained in one domain enhance system-wide performance, integration points developed for one digital worker become available to all, and data collected by one agent enriches context for others.

These effects mean that the value of digital workforces increases superlinearly with scale: the tenth digital worker often delivers more incremental value than the first, reversing the typical pattern of diminishing returns seen with traditional labor.

Total Cost of Ownership (TCO) Considerations

While the fixed-versus-variable cost framework provides a useful starting point, responsible financial analysis must go further. A comprehensive total cost of ownership (TCO) approach accounts for all relevant expenses throughout the digital worker lifecycle:

1. **Development phase costs**

- Licensing
- Process analysis and mapping
- Technical requirements specification
- Implementation partner fees
- Internal personnel time
- Training data collection and curation
- Testing and validation resources

2. **Deployment phase costs**

- Infrastructure setup and configuration
- Data preparation
- Integration with existing systems

- Security and compliance validation
- User training and change management
- Initial monitoring and optimization

3. **Operational phase costs**

- Ongoing computation and storage
- Ongoing usage fees to provider(s)
- Model updates and maintenance
- Performance monitoring and quality assurance
- Governance and compliance activities
- Continuous improvement initiatives

4. **Indirect costs**

- Management oversight
- Risk management measures
- Potential disruption during transition
- Process adaptation and redesign

Organizations that overlook these comprehensive costs often encounter unpleasant surprises that undermine the business case for digital workers.

Particularly in early implementations, the "hidden" costs related to organizational adaptation, governance, and change management can exceed direct technical expenses.

Cost Structures Across Different Digital Worker Categories

The economics of digital labor vary significantly across different use cases. Understanding these variations helps organizations prioritize opportunities based on financial impact.

1. **Transactional digital workers**

- High volume, standardized processes
- Extensive integration with existing systems
- Relatively modest reasoning requirements

Examples: Claims processing, data extraction, routine customer queries

Economic profile: High initial integration costs, extremely low per-transaction costs, rapid breakeven at scale

2. **Knowledge-based digital workers**

- Lower volume, more varied activities
- Substantial context and domain knowledge requirements
- Significant reasoning and judgment capabilities

Examples: Research assistance, content creation, analysis tasks

Economic profile: Higher ongoing operational costs, greater cognitive capabilities, longer payback periods

3. **Strategic digital workers**

- Complex, high-value business activities
- Sophisticated decision-making requirements
- Extensive integration with human collaboration

Examples: Market analysis, scenario planning, product development

Economic profile: Highest capability requirements, significant ongoing costs, value driven by business impact rather than labor replacement

Each category presents a distinct economic case and requires appropriate financial analysis. Organizations that apply transactional ROI expectations to knowledge-based or strategic implementations often abandon potentially valuable invest-ments due to misaligned evaluation frameworks.

Productivity Impact Assessment: Measuring the Real Return on Investment

The true economic value of digital workers extends far beyond simple labor cost replacement, encompassing quality improve-ments, capacity expansion, and entirely new capabilities.

Beyond Labor Substitution: Comprehensive Value Creation

Valuing digital workers solely by the labor costs they replace systematically underestimates their economic impact. A com-prehensive assessment must consider multiple value dimen-sions:

Capacity expansion: Digital workers dramatically increase processing volume without proportional increases in cost.

They eliminate backlogs, reduce delays, and absorb sudden spikes in demand without requiring additional resources. They also operate around the clock, extending service availability beyond conventional working hours.

Quality improvements: Digital workers reduce error rates, minimize the need for rework, and ensure standardized out-puts and decisions.

They offer consistent performance regardless of timing or vol-ume and are often more compliant with policies and regula-tions.

Velocity enhancements: Digital workers accelerate opera-tions by reducing cycle times, eliminating delays caused by manual handoffs, and enabling parallel processing of multiple workflow components.

Their scalability allows organizations to respond instantly to fluctuations in demand.

New capabilities: Perhaps most significantly, digital workers unlock entirely new capabilities.

They enable services that would be economically impractical with human labor alone, such as personalized experiences delivered at scale, real-time handling of complex requests, and continuous data processing. Tasks like automated data preparation and system maintenance become efficient and sustainable.

Organizations that focus exclusively on labor cost replacement typically target low-complexity, high-volume tasks (the low-hanging fruit). In doing so, they often overlook the transformative potential of digital workers when applied to more sophisticated use cases.

Performance Metrics and Evaluation Frameworks

Effective economic assessment requires metrics that reflect the full impact of digital workers across multiple dimensions. These metrics should align with the specific value creation mechanisms of each implementation.

Efficiency metrics provide insight into how digital workers improve operational performance. Key indicators include processing time per transaction, total throughput capacity, resource utilization rates, and cost per output.

Quality metrics focus on the accuracy, consistency, and reliability of digital worker outputs. Organizations should track error rates, rework requirements, and the frequency of compliance violations.

Measures such as output consistency across transactions and precision or recall in decision-making processes offer additional insights.

Business impact metrics evaluate outcomes that affect customers and markets. These include customer satisfaction scores, time-to-resolution for inquiries, revenue generated from new digital capabilities, and changes in market share resulting from improved offerings.

Strategic value metrics assess contributions to organizational agility and response time. They capture how digital workers enhance a company's ability to improve innovation capacity, enable new business models, and differentiate in the marketplace.

The most effective evaluation frameworks combine these metrics into balanced scorecards that prevent myopic focus on cost reduction and capture the broader business value digital workers deliver.

Digital Worker Performance Metrics

Operational

Quality Metrics

Ensures accuracy and consistency

Strategic Value Metrics

Enhances organizational agility

Strategic

Efficiency Metrics

Improves operational performance

Business Impact Metrics

Affects customers and markets

Economic Impact Across Organizational Functions

The economic impact of digital workers varies significantly across different organizational functions.

1. **Customer service and support**

- **Value drivers:** Response time reduction, 24/7 availability, consistency of information
- **Typical ROI factors:** Call deflection, reduced escalations, improved satisfaction scores
- **Implementation considerations:** Integration with knowledge bases, customer context management
- **Measurement approach:** Balanced customer experience and efficiency metrics

2. **Finance and accounting**

- **Value drivers:** Accuracy improvement, audit trail creation, compliance assurance
- **Typical ROI factors:** Error reduction, staff redeployment, audit cost reduction
- **Implementation considerations:** System integration complexity, regulatory requirements
- **Measurement approach:** Combined efficiency and compliance metrics

3. **Operations and supply chain**

- **Value drivers:** Process velocity, exception handling speed, coordination improvements
- **Typical ROI factors:** Inventory reduction, improved fill rates, planning accuracy
- **Implementation considerations:** Multiple system integrations, real-time requirements
- **Measurement approach:** End-to-end process metrics rather than local optimizations

4. **Research and development**

- **Value drivers:** Information synthesis, pattern recognition, ideation support
- **Typical ROI factors:** Time-to-insight reduction, exploration breadth, innovation quality
- **Implementation considerations:** Knowledge access requirements, collaboration capabilities
- **Measurement approach:** Innovation pipeline metrics, time-to-market improvements

Each functional area requires tailored economic analysis approaches that reflect its unique value creation mechanisms and implementation challenges.

Hidden Costs and Benefits Often Overlooked

A complete economic assessment requires considering both hidden costs and benefits that standard ROI analyses frequently miss.

Frequently overlooked costs

1. **Integration complexity**: Connecting digital workers to legacy systems is often more challenging and costly than anticipated.
2. **Knowledge maintenance**: Keeping digital workers updated with current information and policies requires ongoing investment.
3. **Exception management**: Handling cases beyond AI capabilities requires sophisticated escalation processes with significant overhead from the cost of additional human interventions.
4. **Organizational adaptability**: Aligning human workflows and responsibilities with digital workers often requires substantial change management.
5. **Governance overhead**: Monitoring, auditing, and ensuring responsible operation add ongoing costs beyond direct operational expenses.

Hidden Costs of Digital Workers

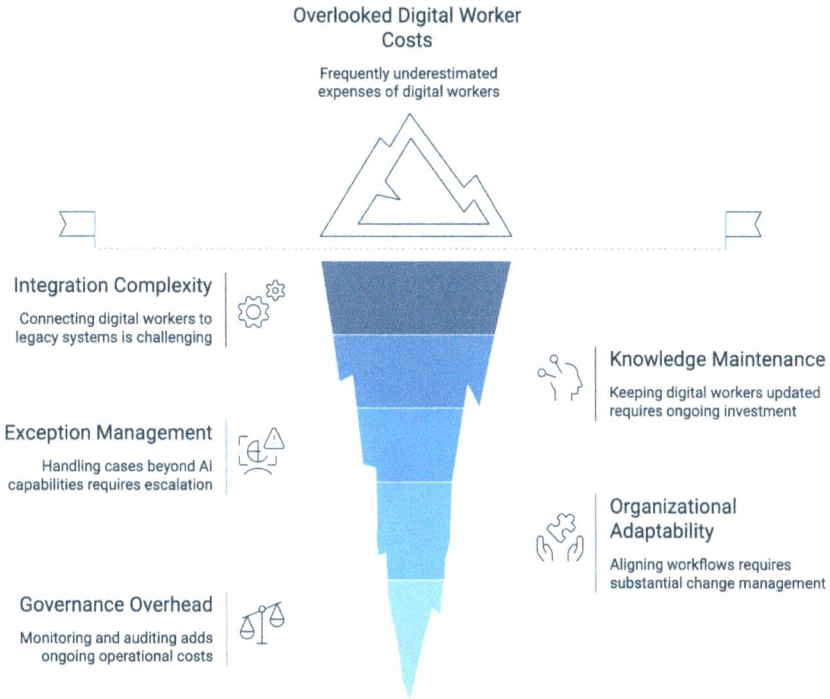

Overlooked Digital Worker Costs

Frequently underestimated expenses of digital workers

Integration Complexity

Connecting digital workers to legacy systems is challenging

Knowledge Maintenance

Keeping digital workers updated requires ongoing investment

Exception Management

Handling cases beyond AI capabilities requires escalation

Organizational Adaptability

Aligning workflows requires substantial change management

Governance Overhead

Monitoring and auditing adds ongoing operational costs

Frequently overlooked benefits

1. **Risk reduction**: Using digital workers significantly reduces operational, compliance, and business continuity risks.

2. **Institutional knowledge capture**: Implementing digital workers often codifies previously tacit knowledge.

3. **Data generation**: Creating structured data about processes through digital workers enables continuous improvement.

4. **Employee experience**: Removing routine tasks from human roles improves satisfaction, reducing turnover and associated costs.

5. **Organizational learning**: Developing digital worker implementation capabilities creates valuable competencies that extend beyond specific projects.

Hidden Benefits of Digital Workers

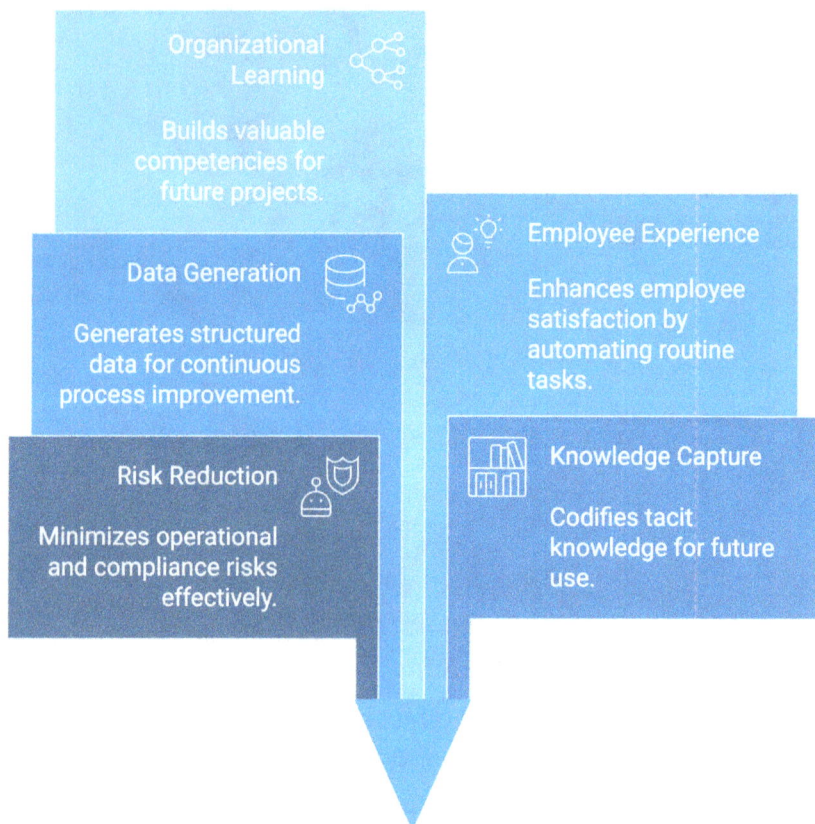

Organizational Learning

Builds valuable competencies for future projects.

Data Generation

Generates structured data for continuous process improvement.

Risk Reduction

Minimizes operational and compliance risks effectively.

Employee Experience

Enhances employee satisfaction by automating routine tasks.

Knowledge Capture

Codifies tacit knowledge for future use.

Organizations that account for these hidden factors develop more realistic business cases and set more appropriate expectations for digital workforce initiatives.

Labor Market Transformations: Broader Economic Implications

Digital workers are reshaping the structure of organizational labor markets, making it essential not only for economic assessment but also for strategic workforce planning and risk management.

Complementarity vs. Substitution Effects on Human Employment

The relationship between digital and human workers is complex, involving both substitution (replacement of human labor) and complementarity (enhancement of human capabilities). The balance between these depends on:

Task characteristics: Routine, rule-based tasks are more likely to be automated entirely, leading to substitution. Complex tasks that require judgment or nuance benefit from complementarity effects.

Creative and strategic activities remain the domain of humans, with digital workers almost exclusively playing a complementary role.

Implementation approach: Automation-focused initiatives prioritize cost reduction and labor replacement. Augmentation strategies aim to improve human productivity and decision-making.

A transformation-focused approach goes even further, using human–AI collaboration to create entirely new capabilities that neither could achieve independently.

Organizational context: Cost-driven organizations often pursue substitution-heavy models to improve margins. Growth-oriented organizations are more likely to adopt complementary approaches.

Innovation-focused companies typically embrace transformative models that create new capabilities as a result of the human-AI collaboration.

Organizations focusing exclusively on substitution effects capture only a fraction of the potential value available from digital workforce implementations. The most substantial economic returns often come from complementarity effects that create new capabilities rather than simply reducing labor costs.

Skill Demand Shifts and Workforce Development Needs

As digital workers assume increasing responsibility for routine tasks, the skill requirements for human workers shift dramatically. This creates both challenges and opportunities for workforce development.

Demand is declining for roles centered around routine information processing, basic customer interactions, standardized analysis, data collection and entry, and rule-based decision-making.

At the same time, there is a growing need for skills such as exception handling, complex problem-solving, emotional intelligence, and interpersonal communication. Creative thinking, strategic judgment, and innovation are becoming increasingly valuable, along with the ability to supervise, fine-tune, and collaborate with digital systems.

Organizations anticipating these shifts should develop proactive workforce transition strategies that maintain business continuity while creating growth paths for employees. Those that fail to address these changes often face simultaneous challenges of redundant skills in some areas and critical shortages in others.

The economic implications of these shifts extend beyond direct labor costs. Companies must also account for the costs of retraining and upskilling, recruiting talent in new capability areas, managing temporary productivity dips during transition, and mitigating turnover from misaligned role expectations.

Forward-thinking organizations align human-digital workforce implementation strategies with talent development initiatives to manage these transitions effectively.

Wage and Employment Effects Across Different Sectors

The economic impact of digital workers varies substantially across different sectors and job categories, creating a complex mosaic of effects rather than uniform patterns.

Most vulnerable sectors

- Administrative support
- Customer service operations
- Data processing and entry
- Basic financial services
- Routine document processing

Most resilient sectors

- Creative and design
- Strategic consulting
- Complex customer relationship management
- Specialized technical work
- Human-centered healthcare

Comparing Sector Resilience and Vulnerability

Routine tasks prone to automation

Creative and strategic roles

Basic customer interaction

Complex customer management

Simple data processing

Specialized technical work

Most Vulnerable Sectors

Most Resilient Sectors

Within these broad patterns, the specific impact depends on several task-level characteristics. Roles dominated by routine work are more likely to be automated, while those requiring emotional intelligence, creative thinking, or nuanced decision-making are less susceptible. The presence of tacit knowledge further insulates some roles from digital substitution.

Organizations should assess their workforce profiles against these factors to develop realistic expectations about the pace and scope of potential transitions. These assessments inform both economic models and change management strategies.

Organizational Requirements

The economic value of digital workers depends not only on their technical capabilities but also on the organization's ability to adapt its structures, processes, and culture to effectively integrate them.

1. **Structural adaptability**

- Redefining reporting relationships and spans of control
- Creating new roles focused on digital worker oversight
- Establishing centers of excellence for implementation support
- Developing cross-functional governance mechanisms

2. **Process adaptability**

- Redesigning workflows to leverage digital worker capabilities
- Establishing handoff protocols between human and digital workers
- Creating exception handling and escalation paths
- Implementing monitoring and continuous improvement cycles

3. **Cultural adaptability**

- Building trust in digital worker capabilities
- Establishing appropriate reliance and verification patterns
- Developing collaborative mindsets for human-digital teams
- Maintaining focus on value creation rather than labor replacement

Organizations investing adequately in these adaptability dimensions typically achieve significantly higher returns on their digital workforce investments. Those that focus exclusively on technical implementation often encounter resistance, underutilization, and missed value opportunities.

Organizational Adaptation for Agentic AI

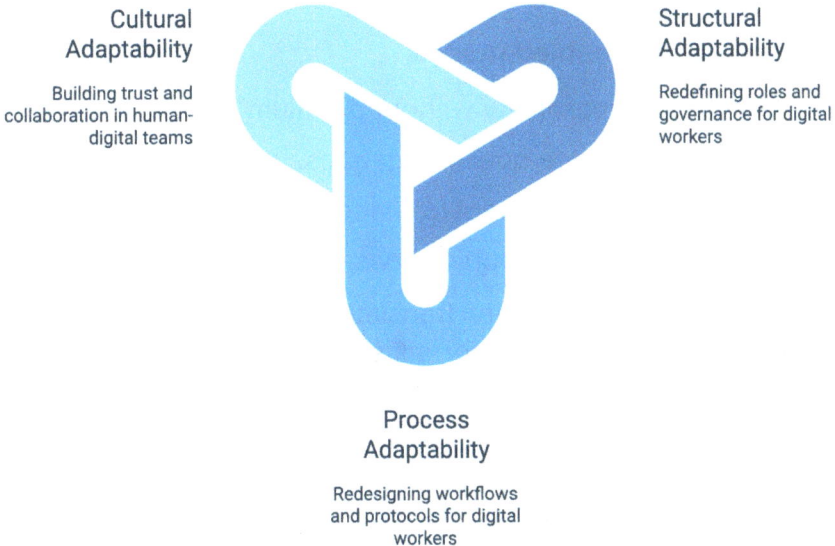

Cultural
Adaptability

Building trust and
collaboration in human-
digital teams

Structural
Adaptability

Redefining roles and
governance for digital
workers

Process
Adaptability

Redesigning workflows
and protocols for digital
workers

Business Model Innovation: New Value Creation Opportunities

Digital workers enable core business model innovations that create entirely new sources of value. These opportunities often represent the highest potential economic returns, though they typically require more substantial organizational transformation.

Service Expansion Through 24/7/365 Capabilities

Digital workers change the economics of continuous service availability, enabling organizations to expand their offerings in ways previously infeasible.

Traditional service models

- Limited hours of availability
- Significant premium for off-hours service
- Capacity constraints during peak periods
- Inconsistent service quality across shifts

Digital worker-enabled models

- Continuous availability without premium pricing
- Consistent service quality regardless of timing
- Dynamic scaling to meet demand fluctuations
- Elimination of geographic service limitations

This expanded availability translates into tangible business value. It enhances customer satisfaction and loyalty and enables companies to capture demand that was previously unserviceable due to time or location constraints.

In competitive, service-sensitive markets, it becomes a key differentiator. And for global businesses, it facilitates market access without the complexity of establishing a local presence.

Organizations leveraging these capabilities can transform service availability from a cost center to a differentiating advantage, particularly in industries with time-sensitive customer needs.

Pricing Model Innovations Enabled by Marginal Cost Reductions

The high fixed costs but minimal marginal cost structure enables pricing innovations that can unlock new market segments and revenue streams.

Traditional pricing constraints

- Minimum price floors based on labor costs
- Service tiering limited by staffing requirements
- Transaction-based models constrained by processing capacity
- Personalization limited by economic feasibility
- Digital worker-enabled pricing models
- "Freemium" offerings with sustainable economics
- Micro-transaction services that were previously uneconomical
- Outcome-based pricing supported by minimal marginal costs
- Subscription models with unlimited usage options

These pricing innovations offer strategic advantages. They allow organizations to reach customer segments that were previously unprofitable, reduce barriers to initial product or service adoption, and generate more predictable revenue streams. They also enable pricing strategies that better reflect actual value delivered to customers.

The most successful implementations often combine new pricing approaches with expanded service capabilities to create compelling new value propositions.

New Product Categories Made Possible by Digital Workers

Beyond improving or expanding existing offerings, digital workers enable entirely new product and service categories.

1. **Hyper-personalized services**

- Customization at a scale impossible with human delivery
- Real-time adaptation to individual preferences
- Contextually aware interactions across touchpoints
- Personalized recommendations integrating multiple data sources

2. **Intelligent products**

- Products with embedded advisory capabilities
- Self-managing systems requiring minimal user intervention
- Adaptive functionality based on usage patterns
- Continuous improvement through usage data analysis

3. **Knowledge synthesis services**

- Integration of vast information resources
- Custom analysis tailored to specific questions
- Real-time information updates and verification
- Multi-source research impossible through manual methods

These new offerings often create their own economic ecosystems, with value creation models different from traditional products and services. Organizations that recognize these possibilities early can establish dominant positions in emerging categories.

Competitive Dynamics and First-Mover Advantages

The economics of digital workforces creates significant potential for first-mover advantages that stem from several mechanisms.

Data accumulation: Earlier implementations gather more training and optimization data, and improved performance creates virtuous cycles of increased adoption.

Customer interaction data further enhances personalization capabilities, while process outcome data fuels continuous improvement and optimization.

Ecosystem development: Early adopters develop integration capabilities that make future deployments faster and more efficient.

As digital workforce platforms mature, they often build strong partner networks, including complementary service providers that enhance their overall offering. Early players also help define standards, protocols, or platform norms.

Organizational learning: Experience with early implementation builds in-house expertise, accelerating future rollouts. These organizations develop risk management capabilities and governance frameworks that enable responsible scaling.

They begin to build a culture of adaptability that allows digital workers to be integrated more effectively into evolving workflows.

Organizations that recognize these dynamic effects can develop implementation strategies that prioritize capability building and competitive positioning alongside immediate ROI considerations.

Conclusion: The New Economics of Work

The economics of digital labor is a major shift in how organizations create value, manage costs, and develop competitive advantages. This shift extends far beyond simple labor cost arbitrage to encompass new service models, pricing innovations, and entirely new product categories.

Organizations hoping to capture the full economic potential of digital workers must:

1. **Look beyond simple cost replacement** to identify transformative value opportunities
2. **Develop comprehensive measurement frameworks** that capture multidimensional impact
3. **Anticipate workforce transitions** and proactively manage skill evolution
4. **Invest in organizational adaptability** alongside technical implementation
5. **Explore business model innovations** enabled by new economic structures

As digital workers evolve from narrow task automation to true agency, their economic impact will continue to expand. The organizations that thrive will be those that recognize this evolution not merely as a cost-saving opportunity but as a big shift in how value is created and delivered.

CHAPTER 3:
STRATEGIC WORKFORCE PLANNING IN THE AGE OF AI

TL;DR:

- Successful digital workforce implementation requires systematic approaches to task allocation based on the comparative advantages of human and digital workers.

- Effective capability mapping considers both technical feasibility and business value dimensions to identify high-potential applications.

- Organizational structures must evolve to accommodate digital workers, with new reporting relationships, spans of control, and governance mechanisms.

- As digital workers assume routine and analytical tasks, human skills must evolve toward areas of distinctive human advantage: complex problem framing, emotional intelligence, ethical judgment, and collaborative leadership.

- Organizations that develop integrated human-digital workforce strategies create both greater implementation success and smoother transitions for affected employees.

Hybrid Workforce Management

The emergence of agentic AI requires rethinking workforce planning and management. Organizations must now consider a hybrid workforce that combines human and digital labor, each with distinct capabilities, limitations, and economics.

Digital-Human Workforce Integration: Creating Optimal Combinations

The integration of digital and human workers signals a change in organizational design and management. Rather than replacing human labor, successful organizations are creating integrated teams where each type of worker contributes their unique strengths to achieve outcomes neither could accomplish alone.

Task Allocation Frameworks Based on Comparative Advantage

Effective integration begins with a systematic approach to task allocation based on the comparative advantages of human and digital workers. While the established approach to automation has focused on "routine" tasks and reserved "complex" work for humans, this binary framing oversimplifies the capabilities of today's agentic AI systems.

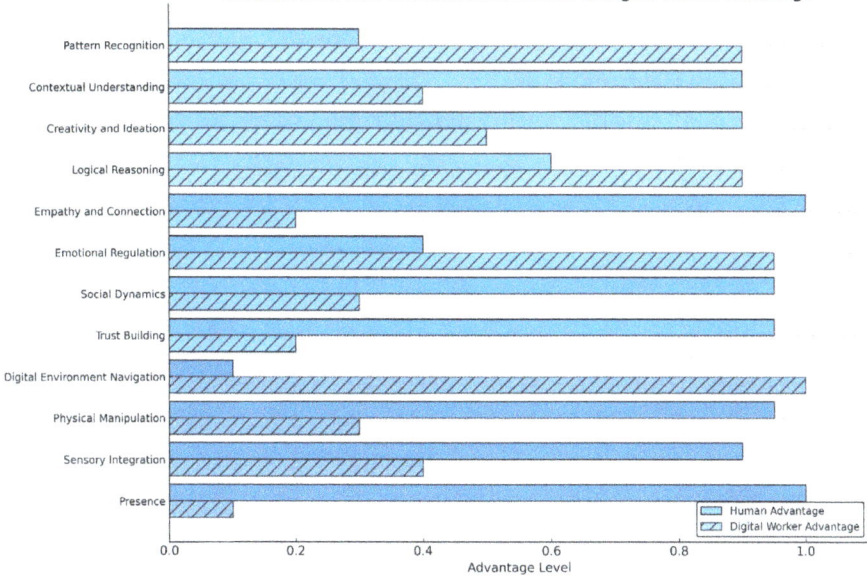

Dimensions of Task Characteristics: Human vs Digital Worker Advantage

A more sophisticated framework considers multiple dimensions of task characteristics:

1. **Cognitive dimension**

- **Pattern recognition**: Digital workers excel at identifying patterns across vast datasets

- **Contextual understanding**: Humans outperform in interpreting ambiguous contexts

- **Creativity and ideation**: Humans lead in generating novel approaches, while digital workers amplify creative processes

- **Logical reasoning**: Digital workers perform well in structured reasoning tasks, while humans excel at intuitive leaps

2. **Emotional dimension**

- **Empathy and connection**: Humans hold clear advantages in genuine emotional understanding

- **Emotional regulation**: Digital workers maintain consistent "emotional" states regardless of context
- **Social dynamics**: Humans excel at navigating complex social environments and power structures
- **Trust building**: Human-to-human connections remain essential in high-trust contexts

3. **Physical dimension**

- **Digital environment navigation**: Digital workers efficiently navigate digital systems and interfaces
- **Physical manipulation**: Humans maintain advantages in dexterity and physical adaptability
- **Sensory integration**: Humans excel at combining multiple sensory inputs in unstructured environments
- **Presence**: Physical human presence remains essential in many contexts.

Task Allocation Based on Comparative Advantage

	Digital Workers	Humans
Pattern Recognition	Excel	Limited
Contextual Understanding	Limited	Outperform
Creativity and Ideation	Amplify	Lead
Logical Reasoning	Perform well	Excel
Empathy and Connection	Limited	Advantage
Emotional Regulation	Consistent	Variable
Social Dynamics	Limited	Excel
Trust Building	Limited	Essential
Digital Navigation	Efficient	Limited
Physical Manipulation	Limited	Advantage
Sensory Integration	Limited	Excel
Presence	Limited	Essential

The most effective task allocation approaches require a detailed understanding of work at the activity level rather than the job level, decomposing roles into their constituent tasks and analyzing each independently.

Collaborative Workflows That Leverage Unique Capabilities

Beyond static task allocation, leading organizations are developing dynamic collaborative workflows where human and digital workers interact continuously, each contributing their unique capabilities at appropriate points in the process.

Most common collaboration models

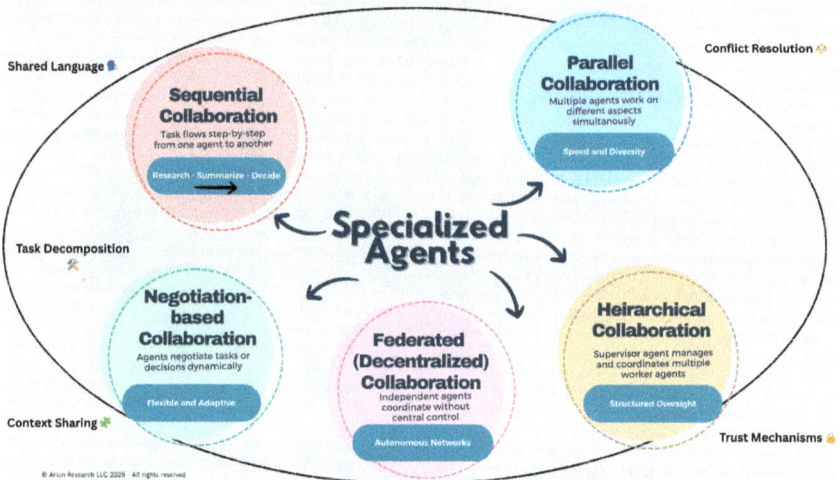

Sequential collaboration: Digital workers handle initial information gathering and synthesis before handing it off to a human to review, validate, and add contextual insights. Digital workers then implement and document decisions while humans provide final approval and accountability.

Parallel collaboration: Digital workers generate alternative analyses while humans develop their own perspectives. Integration points combine insights from both streams, and iterative cycles refine the results by leveraging the strengths of each.

Oversight collaboration: Digital workers execute end-to-end processes with humans intervening only in exceptions and special cases. Supervision structures are put in place to maintain appropriate control, and continuous improvement mechanisms are used to capture insights and refine performance over time.

Negotiation-based collaboration: Agents dynamically negotiate tasks or decisions, allowing for high adaptability in uncertain or rapidly changing environments.

Federated (decentralized) collaboration: Independent agents coordinate without central control, forming autonomous networks.

Hierarchical collaboration: A supervisor agent manages and coordinates multiple worker agents, with structured oversight ensuring alignment and control across tasks.

And in all models, effective collaboration requires shared language, conflict resolution, task decomposition, context sharing, and trust mechanisms.

Collaborative platform integration models

Digital workers are increasingly integrated into existing human collaboration platforms and workflows, eliminating the need for separate interfaces or systems. This integration takes several forms:

Messaging platform integration: Digital workers operate as "members" within Slack, Microsoft Teams, and similar platforms, enabling natural conversational interactions within existing channels and direct messages. These integrations allow digital workers to:

- Respond to direct inquiries in channel conversations
- Monitor discussions for relevant topics requiring intervention
- Provide real-time assistance during virtual meetings
- Access and share resources from connected systems

CRM and workflow integration: Digital workers embedded within platforms like Salesforce, Zoho Corp, or Monday.com augment human decision-making within established business processes. They:

- Provide contextual recommendations during customer interactions
- Automate data entry and enrichment while humans focus on relationships
- Trigger appropriate workflows based on conversation content
- Maintain comprehensive interaction records with minimal human effort

Knowledge management integration: Digital workers connected to knowledge management systems serve as active knowledge brokers, not just passive repositories. They:

- Suggest relevant resources during team discussions
- Keep information current by flagging outdated content
- Connect related conversations across different teams
- Capture and organize insights from ongoing collaboration

Cross-platform orchestration: Advanced digital workers can coordinate activities across multiple systems, serving as integration points between collaboration platforms, operational systems, and specialized tools. They:

- Maintain context across system boundaries
- Translate requests from conversational to structured formats
- Orchestrate multi-step processes spanning multiple systems
- Provide unified interfaces for complex cross-system workflows

These platform integrations reduce friction in human-digital collaboration by eliminating context switching and enabling interactions within familiar environments. They build on existing digital adoption and extend platform capabilities without requiring users to learn entirely new interfaces or workflows.

These collaborative workflows require thoughtful design of interaction points, clear role delineation, and effective information sharing. They transform the human-digital relationship from one of replacement to one of augmentation and enhancement.

Change Management for Successful Integration

The integration of digital workers is a change in organizational structures, individual roles, and established processes. Successful implementation requires a comprehensive change management approach tailored to the unique challenges of digital-human integration. The key elements of effective change management include:

Vision development and communication: Leaders must clearly articulate the purpose and expected benefits of the digital workforce, emphasizing augmentation rather than replacement.

This vision should be tied to broader organizational objectives, with realistic expectations set around timelines and the pace of transition.

Stakeholder engagement and management: Involving affected teams early in the design process, maintaining transparency about role impacts, and creating regular opportunities for feedback builds trust and reduces resistance. Visible executive sponsorship and participation further reinforce the importance of the initiative.

Role evolution support: Organizations should communicate how roles will change, provide targeted training for new responsibilities, and outline future career pathways for affected employees. Recognizing and rewarding adaptability can motivate individuals to embrace new ways of working.

Implementation pacing and sequencing: Gradual changes that allow time for adaptation, prioritizing early wins, and maintaining flexibility to adjust based on feedback significantly improve outcomes. The right balance of challenge and support helps build momentum without overwhelming teams.

Organizations that invest adequately in these change management dimensions typically achieve higher success rates than those focusing exclusively on technical implementation.

The human experience of digital worker integration often determines adoption rates, utilization levels, and ultimately the realized value of investments.

Capability Mapping: Identifying High-Value Applications

Strategic workforce planning begins with identifying where digital workers create the most significant value.

Assessment Methodologies for Task Suitability

Determining which tasks and processes are most suitable for digital worker implementation requires multi-dimensional analysis that considers both technical feasibility and business value.

Technical feasibility dimensions

1. **Perception requirements**

- Types and complexity of inputs
- Structured versus unstructured information
- Multimodal processing needs
- Environmental awareness

2. **Reasoning complexity**

- Decision-making complexity and ambiguity

- Number of variables and considerations
- Exception frequency and handling
- Judgment and intuition

3. **Action capabilities**

- Types of outputs and actions
- System integration
- Feedback processing
- Precision and consistency

4. **Learning and adaptation needs**

- Stability versus evolution of the task environment
- Feedback richness and availability
- Performance improvement
- Adaptability frequency
- Business value dimensions

5. **Volume and scale**

- Transaction or activity frequency
- Consistency across instances
- Geographic and temporal distribution
- Scaling limitations of current approaches

6. **Value creation mechanisms**

- Cost reduction opportunities
- Quality and consistency improvements
- Speed and throughput enhancements
- New capability creation

7. **Strategic importance**

- Customer experience impacts
- Competitive differentiation potential
- Core versus supporting process status
- Innovation enablement

8. **Implementation complexity**

- Integration with existing systems
- Data availability and quality
- Process standardization levels
- Stakeholder complexity and concerns

A scoring framework combining these dimensions enables systematic comparison and evaluation, tailored to each organization's context and strategic priorities.

Prioritization Frameworks for Implementation Sequencing

Once potential implementation opportunities are assessed, organizations must determine the optimal sequence for digital workforce development. This sequencing should balance multiple considerations beyond simple ROI calculations. Value-driven sequencing factors include:

Financial return trajectory includes investment requirements, timing of cost savings or revenue enhancement, payback period and ROI characteristics, and alignment with available budget and funding.

Capability-building path includes how each initiative contributes to the development of technical foundations, enables organizational learning, creates reusable components, and supports the sequential development of skills and experience.

Change management considerations include the organization's readiness by area, its capacity to absorb change, opportunities for early wins that build momentum, and the specific concerns or priorities of stakeholders.

Risk management approach includes implementation complexity and uncertainty, potential disruptions to the business and its impact, available fallback and mitigation options, and any relevant regulatory or compliance issues.

These models typically evolve as organizations gain implementation experience and better understand their specific success factors and challenges.

Risk-Reward Evaluation Models

Digital workforce implementations involve significant potential rewards and substantial risks. Comprehensive evaluation frameworks must address both dimensions and their interrelationships. Common risk categories include:

1. **Technical risks**

- Performance shortfalls against expectations
- Integration challenges with existing systems
- Data quality and availability
- Scalability and reliability

2. **Operational risks**

- Process disruption during transition
- Exception handling and fallback mechanism failures
- Governance and oversight gaps
- Security and privacy vulnerabilities

3. **Organizational risks**

- Resistance to adoption and utilization

- Skill gap emergence during transition
- Role and responsibility confusion
- Cultural misalignment and rejection

4. **Strategic risks**

- Misalignment with changing business priorities
- Competitive leapfrogging during implementation
- Customer acceptance and satisfaction
- Regulatory and compliance challenges

Risk-Reward Evaluation in Digital Workforce Implementation

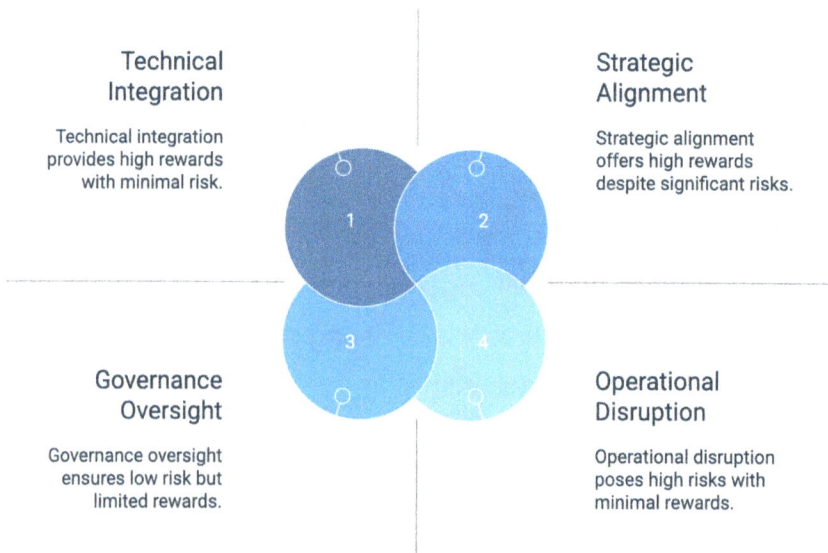

Technical Integration

Technical integration provides high rewards with minimal risk.

Strategic Alignment

Strategic alignment offers high rewards despite significant risks.

Governance Oversight

Governance oversight ensures low risk but limited rewards.

Operational Disruption

Operational disruption poses high risks with minimal rewards.

Effective risk-reward models explicitly link these risk categories to value creation mechanisms, enabling organizations to make informed trade-offs and develop appropriate mitigation strategies.

These models should be dynamic, evolving as implementation proceeds and new information becomes available.

Journey-Based Implementation Planning

Rather than viewing digital workforce initiatives as standalone projects, leading organizations develop integrated maps that connect individual implementations into coherent capability development paths. The key elements of journey-based planning include:

Foundation building stages include the development of technical infrastructure, improvements to data strategy and quality, establishment of governance frameworks, and initial investments in skill development and team formation.

Capability development sequences include logical progression from simple to complex implementations. Along the way, they create reusable components, accumulate knowledge and experience, and establish repeatable patterns that lead to scaled impact.

Organizational adaptation synchronization involves aligning cultural evolution with technical capability growth. This involves carefully coordinating changes to roles and responsibilities, sequencing process redesigns, and aligning skill development and talent acquisition with the organization's transformation pace.

Value realization waypoints include clear milestones for demonstrable business impact, ROI measurement and validation checkpoints, and triggers for scaling or course correction.

Journey-based planning brings coherence to otherwise disconnected initiatives, maximizing synergies, building momentum, and maintaining strategic alignment.

Organizational Design: Restructuring for the Digital Workforce Era

Introducing digital workers requires a rethink of organizational design. Traditional structures built for human workforces must evolve to accommodate the unique characteristics of digital workers and enable effective human-digital collaboration.

New Organizational Structures to Accommodate Digital Workers

While digital workers can be implemented within existing structures, organizations that redesign their architecture to fully leverage digital capabilities typically achieve significantly greater value. This redesign involves several dimensions:

Functional integration model

1. **Distributed integration**

- Digital workers embedded within existing functional teams
- Operational management by function leaders
- Technical support from central resource teams

Advantages: Close alignment with business needs, rapid adaptation

Challenges: Inconsistent implementation, duplicate development

2. **Center of excellence model**

- Centralized digital workforce development and management
- Deployment to functional areas based on prioritization
- Specialized expertise concentration

Advantages: Standardization, skill development, resource optimization

Challenges: Potential disconnection from business needs, prioritization conflicts

3. **Hybrid or federated model**

- Central architecture and governance standards
- Distributed implementation and management
- Shared resources and expertise

Advantages: Balanced standardization and flexibility, scale economies with business alignment

Challenges: Coordination complexity, clear accountability

The optimal model depends on organizational size, digital workforce scale, implementation maturity, and strategic priorities. Many organizations evolve through these models as their digital workforce expands and matures.

Reporting Relationships and Management Spans

Digital workers alter how management functions. Traditional reporting structures and spans of control must be redefined. The key considerations in reporting design include:

Technical vs. operational management: Organizations must clearly separate responsibilities for maintaining the technical performance of digital workers from the oversight of their operational effectiveness.

This includes defining decision rights related to performance optimization, establishing distinct escalation pathways for different issue types, and balancing the pursuit of technical excellence with alignment to business objectives.

Spans of control recalibration: The oversight needs of digital workers differ from those of human employees. Supervisory roles may evolve to encompass both technical and operational responsibilities, demanding new skill sets and management capacity. Performance management for digital workers also follows different approaches.

Governance integration: Digital workers must be represented within broader governance structures. This includes creating accountability frameworks for digital worker outcomes, managing risk and regulatory responsibilities, and embedding ethical oversight to ensure value alignment.

Organizations must develop these structures thoughtfully, not merely extend existing human-centered approaches. The most effective designs account for the distinct nature of digital workers while maintaining clear accountability for outcomes.

Role Evolution and Job Redesign Methodologies

As digital workers take on specific tasks and activities, human roles must evolve accordingly. This evolution requires systematic job redesign that leverages new opportunities rather than simply eliminating positions. Effective job redesign approaches include:

Job Redesign Approaches

Characteristic	Task Analysis	Value-Added Focus	Collaboration Models	Career Pathway
Task Definition	Decomposition into tasks	Identify high-value contributions	Delineate responsibilities	Creation of new roles
Digital Integration	Assess tasks for digital workers	Delegate supporting activities	Design communication mechanisms	Skill development opportunities
Human Role	Recombine human tasks	Concentrate on human skills	Exception handling protocols	Transition support mechanisms
Improvement	Add new responsibilities	Enhance strategic components	Continuous learning mechanisms	Recognition and reward alignment

1. **Task-level analysis and reassembly**

- Decomposition of current roles into constituent tasks
- Assessment of tasks appropriate for digital workers
- Recombination of remaining human tasks into coherent roles
- Addition of new responsibilities leveraging distinctly human capabilities

2. **Value-added focus shifting**

- Identification of highest-value human contributions
- Delegation of supporting activities to digital workers

- Concentration of human time on judgment, creativity, and relationship dimensions
- Enhancement of strategic versus transactional components

3. **Collaboration models definition**

- Clear delineation of human and digital worker responsibilities
- Design of handoff and communication mechanisms
- Exception handling and escalation protocols
- Continuous improvement and learning mechanisms

4. **Career pathway development**

- Creation of new roles focused on digital worker oversight
- Skill development opportunities for affected employees
- Transition support mechanisms and timing
- Recognition and reward alignment with new contributions

The most successful organizations approach job redesign as an opportunity to enhance human work. This orientation typically produces greater value creation and smoother transitions.

Skills and Competency Evolution: Preparing the Human Workforce

As digital workers assume increasing responsibility for routine and analytically complex tasks, the skill requirements for human workers evolve significantly. Organizations must proactively address these shifts through a comprehensive talent strategy.

Critical Human Skills in the Age of Agentic AI

Distinctly human skills will become increasingly valuable as AI continues to advance. These skills cluster in several domains:

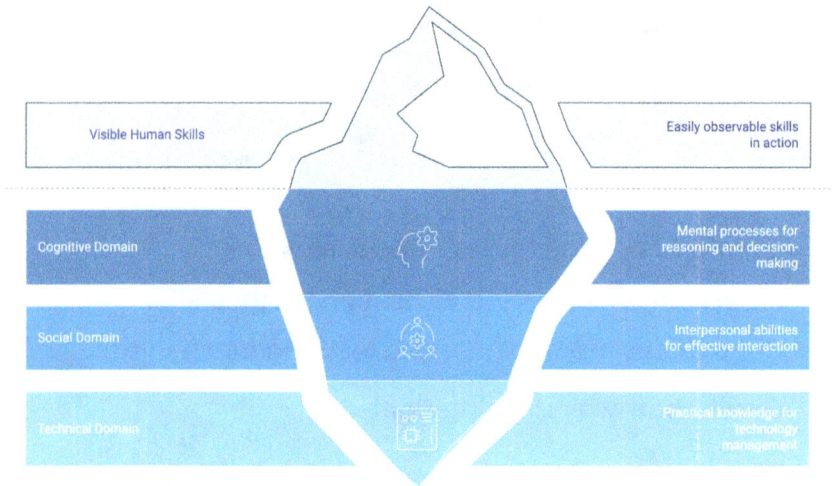

Critical Human Skills in the Age of Agentic AI

Visible Human Skills	Easily observable skills in action
Cognitive Domain	Mental processes for reasoning and decision-making
Social Domain	Interpersonal abilities for effective interaction
Technical Domain	Practical knowledge for technology management

Cognitive domain

1. Complex problem framing

- Identifying the right problems to solve
- Recognizing underlying issues beyond symptoms
- Establishing appropriate context and boundaries
- Determining evaluation criteria and success measures

2. Creative synthesis

- Combining diverse inputs into novel solutions
- Transcending established patterns and approaches
- Applying cross-domain analogies and insights
- Identifying non-obvious connections and implications

3. Ethical judgment

- Navigating value conflicts and trade-offs
- Considering diverse stakeholder perspectives
- Evaluating long-term societal implications

- Aligning decisions with organizational values

4. **Strategic foresight**

- Anticipating emerging trends and developments
- Identifying potential disruptors and opportunities
- Developing robust strategies amid uncertainty
- Balancing short-term and long-term considerations

Social domain

1. **Emotional intelligence**

- Recognizing and responding to emotional states
- Building genuine human connection and trust
- Navigating sensitive situations with empathy
- Creating psychological safety in teams

2. **Collaborative leadership**

- Inspiring and aligning diverse stakeholders
- Facilitating productive group dynamics
- Fostering innovation through inclusive approaches
- Building consensus around complex decisions

3. **Influence and persuasion**

- Articulating compelling narratives and visions
- Adapting communication to different audiences
- Building coalitions around change initiatives
- Overcoming resistance through understanding

4. **Cross-cultural fluency**

- Navigating diverse cultural contexts effectively
- Recognizing and respecting different perspectives

- Adapting approaches across cultural boundaries
- Building inclusion across diverse populations

Technical domain

1. **Digital worker oversight**

- Understanding AI capabilities and limitations
- Evaluating digital worker performance effectively
- Identifying improvement opportunities
- Managing exceptions and edge cases

2. **Human-AI collaboration**

- Working effectively with digital colleagues
- Providing appropriate guidance and feedback
- Understanding when to trust or verify outputs
- Communicating effectively with AI systems

3. **Complex system design**

- Architecting integrated human-digital systems
- Designing effective collaboration mechanisms
- Establishing appropriate governance frameworks
- Creating adaptive learning mechanisms

4. **Data literacy and ethics**

- Understanding data sources and limitations
- Recognizing potential biases and risks
- Making ethical decisions about data use
- Translating insights into business actions

Organizations that proactively develop these skills create both competitive advantage through superior human capabilities and greater resilience amid technological change.

Upskilling and Reskilling Strategies

As skill requirements evolve, organizations must adopt comprehensive approaches to develop new capabilities within their workforce. Effective strategies combine multiple learning modalities and clear development pathways. Key components of effective skill development include:

Comprehensive skill gap analysis includes assessing the current capability baseline, projecting future role-specific requirements, prioritizing based on business impact, and identifying individual development needs.

Multimodal learning approaches include formal training programs for foundational knowledge, project-based learning for hands-on experience, coaching and mentoring for contextual skills, and peer learning communities to support continuous development and knowledge sharing.

Progressive development pathways include defining clear skill levels and progression markers, breaking down capability development into manageable steps, linking progress to career advancement opportunities, and recognizing achievements to encourage skill acquisition.

Measurement and accountability approaches include tracking skill development with clear metrics, holding managers accountable for building team capabilities, encouraging individual responsibility for learning, and connecting learning progress to performance management systems.

Organizations that invest in these approaches achieve greater digital workforce implementation success and higher employee retention during transitions.

Career Path Evolution and Talent Development Approaches

Beyond immediate skill development, organizations must rethink career pathways to reflect the changing nature of work in the age of agentic AI. This involves modifying existing career tracks and creating entirely new paths, such as:

1. **Technical to strategic progression**

- Evolution from task execution to problem framing
- Increasing focus on exception handling and judgment
- Greater emphasis on innovation and improvement
- Integration of technical and business perspectives

2. **New specialization tracks**

- Human-AI collaboration specialists
- Digital worker supervisors and trainers
- Process redesign and optimization experts
- Implementation and change management leaders

3. **Lateral development emphasis**

- Greater value in cross-functional experience
- Parallel skill development across multiple domains
- Recognition of diverse capability combinations
- Customized development paths based on strengths

4. **Lifelong learning integration**

- Continuous development expectations
- Regular reskilling opportunity
- Learning capability as a core competency
- Innovation and adaptability rewards

Organizations that proactively evolve these career structures send powerful signals about future direction and provide crucial support for workforce transitions.

Talent Acquisition and External Workforce Strategies

While internal development is a primary approach to capability building, organizations must also consider talent acquisition and alternative workforce models to access critical skills. Evolving talent acquisition approaches include:

New skill prioritization includes shifting from purely technical execution to more creative and strategic capabilities. There is an increased emphasis on adaptability, a learning mindset, and the value of diverse experiences and perspectives, along with strong collaboration and communication abilities.

Alternative sourcing models include the strategic use of a contingent workforce for specialized skills, ecosystem partnerships to access capabilities, collaborations with educational institutions to build future talent pipelines, and even the acquisition of specialized teams.

Talent brand evolution includes positioning digital transformation around augmentation rather than replacement. This involves emphasizing opportunities for meaningful human contribution, highlighting investment in employee development, and demonstrating ethical, responsible approaches to AI.

Organizations that effectively balance internal development with external acquisition create immediate capability access and sustainable long-term talent strategies.

Conclusion: The New Frontier of Workforce Strategy

Strategic workforce planning in the age of AI represents a major rethinking of how organizations approach their most valuable resources.

Rather than viewing digital workers as simple automation tools, forward-thinking organizations recognize them as a new category of contributors requiring thoughtful integration, organizational adaptability, and human capability evolution.

The organizations that will thrive amid this transition share several characteristics:

1. **Integrated workforce vision**: They develop comprehensive strategies that address both digital and human workforce evolution in a coordinated fashion.

2. **Systematic implementation approach**: They employ structured methodologies for capability mapping, prioritization, and risk management rather than opportunistic automation.

3. **Organizational design focus**: They recognize that maximizing value requires rethinking structures, roles, and management approaches rather than simply adding technology.

4. **Human-centered orientation**: They emphasize augmentation and enhancement of human capabilities rather than replacement, creating greater value and smoother transitions.

5. **Forward-looking skill development**: They proactively address emerging skill requirements through comprehensive development programs and career path evolution.

As agentic AI capabilities continue to advance, the gap between organizations that approach workforce planning strategically versus tactically will widen. Those that recognize the depth of this transition and invest accordingly will create sustainable competitive advantages through the unique capabilities of their integrated human-digital workforces.

PART
TWO
IMPLEMENTATION AND OPERATIONS

CHAPTER 4:
DESIGNING EFFECTIVE DIGITAL WORKERS

TL;DR:

- Goal alignment is the foundation of effective digital worker design, requiring careful objective specification, constraint definition, and value alignment methodologies.

- Well-designed system architecture enables reliability, integration, and evolution through modularity, appropriate fallback mechanisms, and thoughtful human-in-the-loop approaches.

- Interaction design impacts adoption and effectiveness, with communication protocols, trust-building elements, and feedback mechanisms all playing critical roles.

- The choice between specialization and generalization involves important tradeoffs that should be evaluated based on task characteristics, organizational context, and technical capabilities.

- Successful digital worker design bridges technical and human considerations, creating systems that not only perform effectively but also adapt and evolve over time.

Moving Beyond Simple Automation

The transition from conceptualizing digital workers to implementing them requires careful design considerations that go far beyond simple automation. Creating effective agentic AI systems demands thoughtful approaches to goal alignment, system architecture, interaction design, and capability scope.

Goal Alignment Methodologies: Ensuring Digital Workers Pursue Intended Outcomes

Perhaps the most critical aspect of digital worker design is ensuring these systems reliably pursue the outcomes their human designers and operators intend. As agentic AI becomes more autonomous and capable, aligning their behavior with organizational goals becomes increasingly complex and important.

Objective Specification Techniques and Pitfalls

Clearly specifying what digital workers should accomplish is surprisingly challenging. Even seemingly straightforward objectives may result in unexpected behaviors when implemented in agentic systems with significant autonomy.

Effective objective specification approaches

Outcome-based specification focuses on defining the desired results rather than prescribing specific methods. It involves setting clear success criteria that can be objectively evaluated and establishing explicit timeframes and conditions under which the objective should be achieved.

Example: "Schedule meetings with all team members within the next two weeks, optimizing for their stated availability preferences and minimizing schedule disruption."

Multi-dimensional objectives recognize that the most valuable goals often involve multiple considerations. These objectives provide explicit prioritization among competing dimensions and aim to balance both quantitative and qualitative success measures.

Example: "Respond to customer inquiries with priority given to (1) accuracy of information, (2) compliance with policies, (3) customer satisfaction, and (4) fast response time."

Iterative refinement processes involve setting an initial objective and then reviewing and adjusting it based on performance and feedback. This approach includes testing with diverse scenarios to identify potential issues, incorporating stakeholder feedback, and remaining open to the continuous evolution of the objective based on operational experience.

Example: For AI safety research, initial objective: "Make AI systems safer." 1. First refinement: "Reduce harmful outputs from language models." 2. Second refinement: "Decrease toxic language generation by 90% while maintaining helpfulness scores above baseline." 3. Third refinement: "Achieve toxicity reduction through constitutional AI training with human feedback loops, measured against established benchmarks."

Objective Specification Approaches

	Outcome-Based	Multi-Dimensional	Iterative Refinement
Definition	Desired results defined	Multiple considerations recognized	Objective adjusted based on feedback
Success Criteria	Objectively evaluated	Explicit prioritization provided	Testing with diverse scenarios
Example Focus	Setting clear success criteria	Balancing quantitative and qualitative measures	Continuous evolution of the objective

Common objective specification pitfalls

Over-specification occurs when objectives are defined with excessive detail that constrains potentially valuable approaches.

These objectives often focus on prescribing methods rather than outcomes, placing unnecessary limitations on digital worker discretion.

Example: Dictating specific scripts or workflows rather than desired outcomes.

Under-specification refers to objectives that are vague or ambiguous, making them open to multiple interpretations. These lack necessary constraints or guardrails on acceptable approaches and fail to provide sufficient clarity on what success looks like.

Example: "Improve customer satisfaction" without specific metrics or considerations.

Metric fixation is the overemphasis on aspects of performance that are easy to measure, often at the expense of important qualitative dimensions. This can lead to the creation of contrary incentives or unintended behaviors.

Example: Focusing solely on response time, potentially sacrificing accuracy.

Conflicting objectives arise when contradictory goals are set without clear prioritization. These scenarios often attempt to optimize across multiple dimensions simultaneously, which may be impossible, and provide inconsistent guidance that can create decision paralysis.

Example: Simultaneously requiring minimum cost, maximum quality, and fastest delivery without trade-off guidance.

Achieving Effective Objective Specification

Conflicting Objectives

Contradictory goals without clear prioritization

Over-specification

Objectives with excessive detail that limit flexibility

Metric Fixation

Overemphasis on measurable aspects at the expense of qualitative dimensions

Under-specification

Vague objectives lacking necessary constraints

Organizations that develop strong capabilities in specifying and refining objectives typically achieve significantly better alignment between intended and actual digital worker behavior.

This capability becomes increasingly important as digital workers gain greater autonomy and responsibility.

Constraint Definition and Enforcement

Effective design requires a clear definition of what digital workers should not do. Constraints establish boundaries to prevent harmful or unintended actions.

Key constraint categories

Ethical boundaries

- Prohibiting harmful, deceptive, or manipulative behaviors
- Requiring transparency in certain contexts
- Limiting autonomous decision-making in sensitive areas
- Respecting human autonomy and dignity

6. Operational guardrails

- Limiting resource usage (computational, financial, time)
- Requiring authorization for certain actions
- Restricting access to sensitive systems or data
- Mandating human approval for high-impact decisions

7. Legal and compliance requirements

- Adhering to relevant regulations and policies
- Obligating documentation and record-keeping
- Requiring privacy and data protection
- Complying with industry-specific standards

8. Safety mechanisms

- Preventing irreversible or high-risk actions without verification
- Monitoring requirements for uncertain situations
- Triggering circuit breakers for unexpected behavior patterns
- Initiating fallback protocols when operating outside normal parameters

Key Constraint Categories

	Ethical	Operational	Legal/Compliance	Safety
Description	Prevents harmful behavior	Limits resource consumption	Follows regulations and policies	Prevents high-risk actions

Constraint implementation approaches

Hard constraints enforce absolute limitations at the system level. These are implemented through access controls that prevent the violation of specified boundaries and include technical safeguards that cannot be overridden.

Example: Preventing access to certain databases or system capabilities.

Soft constraints offer guidance rather than strict enforcement. They are integrated into agent decision-making as weighted considerations that influence choices without completely restricting them. This allows for contextual application depending on circumstances.

Example: Preference for certain approaches unless specific conditions apply.

Oversight mechanisms ensure control through human review for certain actions. These mechanisms monitor systems to detect potential violations and include escalation protocols to handle boundary cases. They are also subject to continuous evaluation to ensure constraint effectiveness over time.

The most effective constraint frameworks balance protection against unintended behaviors with sufficient flexibility for digital workers to accomplish their objectives in diverse and changing circumstances.

Value Alignment Approaches and Verification Methods

Truly effective digital workers must align with the broader values and principles of the organizations they serve. This deeper alignment enables appropriate action even in novel situations not explicitly covered by specific instructions.

Key value alignment methodologies

1. **Explicit value articulation**

- Clear statement of organizational values
- Translation of abstract values into concrete guidelines
- Prioritization framework for value conflicts

- Regular review and refinement based on emerging scenarios

2. **Value demonstration through examples**

- Case-based learning from historical decisions
- Scenario training with expert-generated responses
- Comparison-based preference learning
- Continuous refinement through operational feedback

3. **Stakeholder involvement**

- Diverse input during value definition processes
- Multiple perspective consideration in training and evaluation
- Regular review of alignment with different stakeholder needs
- Feedback mechanisms for ongoing adjustment

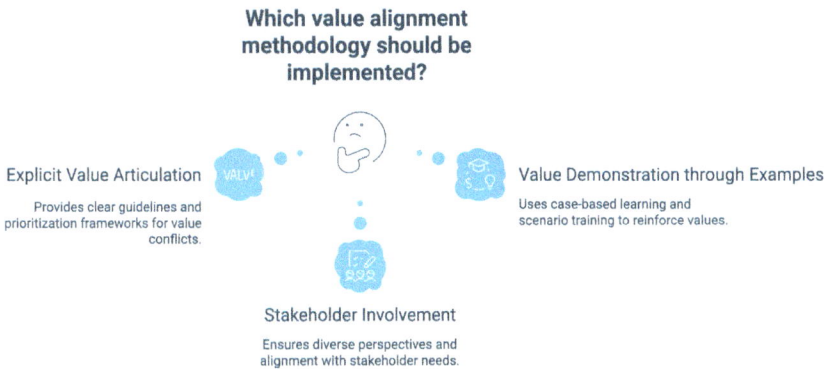

Which value alignment methodology should be implemented?

Explicit Value Articulation
Provides clear guidelines and prioritization frameworks for value conflicts.

Value Demonstration through Examples
Uses case-based learning and scenario training to reinforce values.

Stakeholder Involvement
Ensures diverse perspectives and alignment with stakeholder needs.

Verification approaches

Scenario testing involves evaluating system performance across diverse potential situations. This includes exploration of edge cases for boundary conditions, adversarial testing to uncover potential misalignments, and regular reassessment to ensure continued reliability as capabilities evolve.

Monitoring and evaluation focus on the ongoing assessment of decisions against the value framework, implementing anomaly detection to flag potential misalignments, auditing high-impact actions, and using feedback loops to drive continuous improvement.

Transparent reasoning requires explainability for critical decisions. It emphasizes clear articulation of the values influencing decisions, traceability between values and specific actions, and thorough documentation of trade-offs and prioritization logic.

Value alignment is challenging, but it is also the most important aspect for digital workers with significant autonomy. Organizations that invest in robust alignment approaches build digital workforces that can be trusted with increasingly responsible roles.

System Architecture: Building Robust Agentic Systems

The architectural design of digital workers determines their capabilities, limitations, and ultimate effectiveness. Well-designed architecture enables reliable operation, appropriate human collaboration, and ongoing evolution as requirements change and technologies advance.

Component Organization for Scalable Digital Workers

Effective digital workers typically employ modular architectures that separate different functional capabilities, enabling both specialization and reuse across implementations.

Core architectural components

1. **Perception modules**

- Natural language understanding for text inputs
- Document processing and information extraction
- Structured data analysis

- Multimodal input handling
- Context recognition and situational awareness

2. **Reasoning and decision systems**

- Planning and sequencing
- Knowledge retrieval and application
- Logical reasoning for structured problems
- Uncertainty handling and probabilistic reasoning
- Learning and adaptation mechanisms

3. **Action generation components**

- API interaction
- Content generation systems
- Response formulation mechanisms
- Tool usage coordination
- System navigation

4. **Memory and context management**

- Short-term working memory for current tasks
- Long-term knowledge storage
- Episodic memory for interaction history
- Context preservation across sessions
- Knowledge updating mechanisms

5. **Monitoring and control systems**

- Self-assessment
- Performance monitoring
- Error detection and recovery
- Escalation mechanisms
- Feedback processing

Architectural Framework

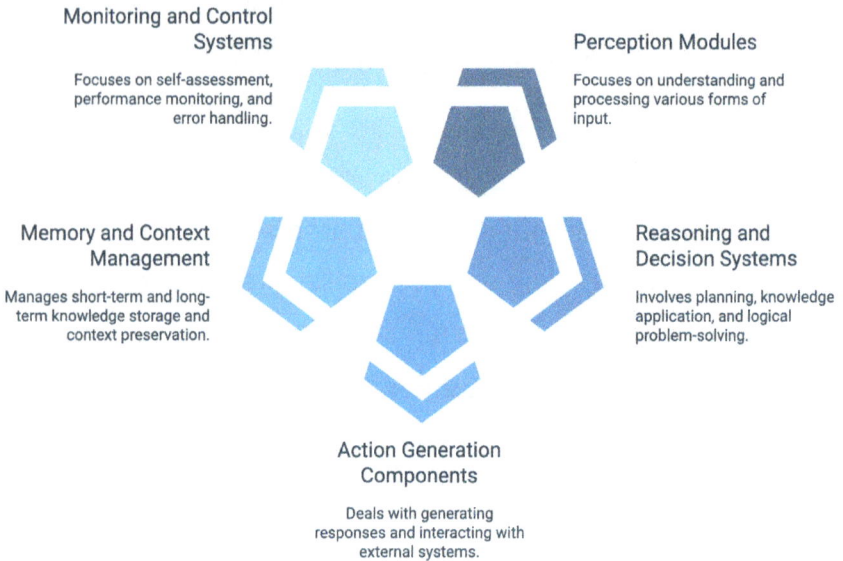

Monitoring and Control Systems

Focuses on self-assessment, performance monitoring, and error handling.

Perception Modules

Focuses on understanding and processing various forms of input.

Memory and Context Management

Manages short-term and long-term knowledge storage and context preservation.

Reasoning and Decision Systems

Involves planning, knowledge application, and logical problem-solving.

Action Generation Components

Deals with generating responses and interacting with external systems.

Architectural design principles

1. Modularity

- Distinct components with well-defined interfaces
- Replacement of individual modules without system redesign
- Separation of concerns across components
- Independent testing and validation

2. Scalability

- Resource allocation based on task requirements
- Ability to handle varying workload volumes
- Performance optimization under different conditions
- Consistent operation across scale changes

3. **Extensibility**

- Clear integration points for new capabilities
- Standardized interfaces for component additions
- Progressive enhancement possibilities
- Future-proofing against evolving requirements

4. **Observability**

- Comprehensive monitoring
- Detailed logging for operational analysis
- Performance metrics collection
- Debugging and diagnostic features

Organizations that invest in thoughtful architectural design typically achieve greater reliability, easier maintenance, and more efficient evolution of their digital workforce capabilities.

Integration with Existing Enterprise Systems

Digital workers rarely operate in isolation. Their effectiveness depends significantly on how well they integrate with an organization's existing technology ecosystem.

Key integration considerations

1. **Authentication and authorization**

- Secure access management to enterprise systems
- Appropriate permission scoping for different functions
- Identity management across integrated systems
- Audit trails for all system interactions

2. **Data access and synchronization**

- Efficient retrieval from existing data sources
- Write-back capabilities where appropriate

- Real-time versus batch processing considerations
- Consistency management across systems

3. **Process integration**

- Handoff mechanisms between digital and existing systems
- Status tracking across system boundaries
- Error handling coordination
- End-to-end process visibility

4. **User experience continuity**

- Consistent interaction models across touchpoints
- Seamless transitions between systems
- Context preservation across boundaries
- Unified status reporting and notifications

Integration architectural patterns

1. **API-based integration**

- REST or GraphQL interfaces for system communication
- Standardized data formats and protocols
- Service-oriented architecture alignment
- Clear versioning and compatibility management

2. **Event-driven architecture**

- Publish-subscribe patterns for system coordination
- Message queues for reliable communication
- Event streams for state synchronization
- Decoupled systems with well-defined interactions

3. **Robotic process automation (RPA) extension**

- Integration with existing RPA infrastructure

- Enhancement of current automations with cognitive capabilities
- Progressive transition from rule-based to agentic approaches
- Leverage of established integration points

4. **Data fabric approaches**

- Unified data access layer across systems
- Consistent information model spanning boundaries
- Metadata management for context preservation
- Knowledge graph integration for relationship awareness

The most successful digital worker implementations create seamless workflows across system boundaries rather than isolated islands of automation.

Model Context Protocol (MCP) in integration

A critical enabler of seamless integration between digital workers and enterprise systems is the Model Context Protocol (MCP). MCP provides a standardized framework for connecting AI agents and digital workflows to a wide range of enterprise applications, data stores, and process engines. By introducing a uniform "language" for interaction, MCP eliminates the complexity and fragmentation of custom integrations, making it easier for organizations to extend automation and intelligence across their digital landscape.

The importance of MCP lies in its ability to:

- **Simplify integration:** MCP standardizes how digital agents interact with various systems, reducing the need for bespoke connectors.
- **Enhance interoperability:** It allows processes and data to flow more freely between legacy platforms, cloud services, and new AI-powered tools.

- **Enable context sharing:** With MCP, digital workers can retain and hand off business context reliably across systems, improving continuity and reducing errors.

- **Future-proof ecosystems:** As new tools and technologies emerge, MCP enables organizations to plug them into existing workflows with minimal friction.

By leveraging MCP, enterprises can create unified, resilient, and context-aware workflows, ensuring that their digital workers operate effectively within complex, evolving technology environments.

Fallback Mechanisms and Human-in-the-Loop Designs

No digital worker, regardless of sophistication, can handle every possible situation effectively. Including appropriate fallback mechanisms and human collaboration capabilities ensures continuity and quality when automated approaches reach their limits.

Fallback mechanism categories

Confidence-based routing involves a digital worker's self-assessment of its ability to handle specific requests. When confidence falls below predefined thresholds, the system automatically escalates the task.

Different handling strategies are applied based on these confidence levels, and the system transparently communicates any uncertainty to the user.

Exception detection and handling refer to the identification of unusual or edge case scenarios. These may be addressed through specialized handling for known exceptions or through graceful degradation when faced with unfamiliar issues.

Additionally, patterns from these exceptions are used to inform improvements in future handling.

Gradual automation approaches use a progressive strategy that moves from supervised to fully autonomous operation.

Initially, only well-understood components are automated, while human involvement is retained in more complex areas.

Over time, autonomous capabilities expand as performance improves and confidence in the system grows.

Seamless human handoffs ensure smooth transitions to human agents when necessary. These handoffs include comprehensive context transfer with relevant information to support continuity.

The system clearly indicates what has been completed and what is pending, and it sets expectations appropriately for all participants.

Human-in-the-loop design patterns

Review and approval workflows involve the digital worker preparing recommendations or content that a human reviews and approves before any implementation.

This setup includes feedback loops for continuous improvement and allows for progressive adjustments in autonomy based on observed performance over time.

Collaborative problem-solving focuses on joint efforts between digital workers and humans for complex cases that require judgment.

It ensures that complementary capabilities are used appropriately, with dynamic allocation of responsibilities based on case characteristics. This collaboration also fosters mutual learning between humans and digital worker systems.

Monitoring and intervention designs include human oversight of autonomous operations, supported by alert mechanisms that signal potential issues.

Humans retain the ability to intervene when necessary, and performance feedback from these interventions helps improve the system over time.

Escalation hierarchies establish tiered responses based on the complexity and impact of situations. Clear routing rules guide the handling of different types of issues, and appropriate information is provided at each escalation level.

Human-in-the-Loop Design Patterns

1. **Digital Worker Preparation**
   Digital worker prepares recommendations

2. **Human Review and Approval**
   Human reviews and approves recommendations

3. **Collaborative Problem-Solving**
   Human and digital worker collaborate on complex cases

4. **Monitoring and Intervention**
   Human monitors and intervenes as needed

5. **Escalation Hierarchy**
   Issues are escalated based on complexity

Resolution tracking and knowledge capture ensure that learning from past escalations is retained and applied.

Effective fallback and human collaboration designs transform potential failure points into opportunities for system improvement.

Interaction Design: Creating Intuitive Human-AI Interfaces

The interface between digital workers and their human colleagues shapes adoption, trust, and ultimate effectiveness. Thoughtful interaction design builds confidence, enables productive collaboration, and supports continuous improvement.

Communication Protocol Design

Clear communication protocols define expectations, enable effective information exchange, and build understanding between human and AI participants.

Key communication protocol elements

Interaction initiation involves clear mechanisms to engage digital workers, such as explicit availability indicators and context-appropriate engagement models. Protocols should also establish transparent expectations for capacity and response times.

Request formulation guidance involves clear instructions on effective query approaches, with templates for common request types, examples of well-structured requests, and progressive guidance based on user experience.

Response formatting involves following a consistent structure across similar responses. They must present an appropriate level of detail for different contexts and clearly distinguish between facts, inferences, and recommendations. The format should be adaptable to user preferences and needs.

Clarification mechanisms involve protocols for handling ambiguous requests, structured formulation for missing information, and confirmation approaches for critical details. The goal is progressive refinement of understanding to support more accurate and meaningful interactions.

Essential Communication Protocols

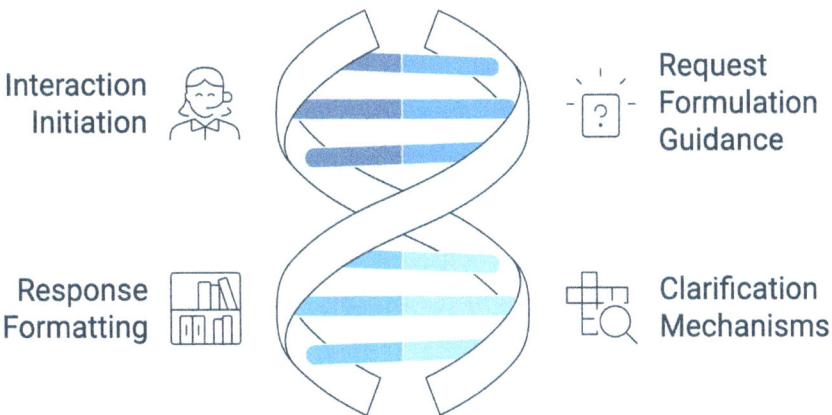

Interaction Initiation

Request Formulation Guidance

Response Formatting

Clarification Mechanisms

Communication style considerations

Transparency calibration: Digital workers should appropriately disclose their limitations and any uncertainty involved in their outputs.

They must offer visibility into their reasoning and information sources and clearly distinguish between different confidence levels. The depth of explanation should be tailored to suit the specific situation.

Persona definition: Communication should maintain a consistent voice and tone that aligns with the organization's culture.

The formality of interactions should match the context, and personality traits should be designed to build trust and engagement. Systems must also adapt their persona to different user preferences and interaction styles.

Cultural alignment: Protocols must reflect the organization's communication norms and show sensitivity to industry-specific conventions.

Terminology and jargon should be used appropriately, with a strong alignment to the brand's voice and values.

The most effective communication protocols evolve through user feedback and interaction patterns, continuously enhancing collaboration effectiveness.

Trust-Building Interface Elements

Trust is foundational to human-AI collaboration. Well-designed interfaces incorporate specific elements that build and maintain trust across interactions.

Trust-building design principles

Appropriate transparency: Users must be able to see what the system can and cannot do, understand the sources behind the information presented, and receive clear explanations of reasoning for recommendations.

It's also important to disclose uncertainty and communicate varying confidence levels appropriately.

Predictable behavior: Digital workers must respond consistently to similar inputs, showing clear indicators of processing status, and performing reliably within stated parameters. Interfaces should also explicitly notify users of changes or updates.

Competence demonstration: Digital workers must showcase capabilities progressively through interaction. This includes effectively handling core use cases, gracefully managing edge cases, and showing continuous performance improvement over time.

Appropriate autonomy: There must be a clear delineation of decision authority, with explicit user confirmations for consequential actions. The system should follow predictable escalation patterns and respect human oversight and direction.

Interface trust elements

1. **Confidence indicators**

- Visual or textual signals of certainty levels
- Clear differentiation between high and low confidence responses
- Appropriate qualifiers in uncertain situations
- Alternative option presentation when confidence is low

2. **Information source visibility**

- Clear attribution of data sources
- Links to reference materials, where appropriate
- Distinction between retrieved and inferred information
- Update recency indicators for time-sensitive data

3. **Reasoning transparency**

- Explicit sharing of decision factors
- Visibility into prioritization and trade-offs
- Option comparison when alternatives exist
- Appropriate detail level based on decision impact

4. **Error recovery mechanisms**

- Graceful handling of mistakes
- Easy correction capabilities
- Learning from error patterns
- Transparent acknowledgment of limitations

Interface Trust Elements

Confidence indicators

Visual or textual signals of certainty levels. Clear differentiation between high and low confidence responses.

Clear attribution of data sources. Links to reference materials, where appropriate.

Information source visibility

Reasoning transparency

Explicit sharing of decision factors. Visibility into prioritization and trade-offs.

Graceful handling of mistakes. Easy correction capabilities. Learning from error patterns.

Error recovery mechanisms

Organizations that prioritize trust-building design elements typically achieve higher adoption rates, stronger collaboration, and greater value realization from their digital workforce implementations.

Feedback Mechanisms and Continuous Improvement

Effective systems incorporate robust feedback mechanisms that enable continuous improvement based on operational experience.

Key feedback system elements

1. **User feedback collection**

- Frictionless mechanisms for input submission
- Structured feedback for common improvement areas
- Open-ended options for novel suggestions
- Contextual collection at relevant interaction points

2. **Performance monitoring**

- Key metric tracking across interaction dimensions
- Anomaly detection for potential issues
- Trend analysis for gradual shifts
- Comparative assessment across similar tasks

3. **Outcome tracking**

- End-to-end process success measurement
- Value realization assessment
- Unintended consequence identification
- Long-term impact evaluation

4. **Learning integration**

- Systematic incorporation of feedback into improvement

- Prioritization frameworks for enhancement focus
- A/B testing for alternative approaches
- Regular capability enhancement releases

Feedback loop implementation

Real-time adaptation involves the immediate application of user corrections, dynamic adjustments in response to interaction patterns, and session-level personalization based on observed behavior.

This ensures that the system remains responsive and contextually aware during active use.

Periodic enhancement cycles involve scheduled analysis of aggregated feedback, systematic prioritization of improvement areas, and regular capability enhancement releases.

These cycles are accompanied by transparent communication with users about what has changed and why.

Collaborative improvement includes user involvement in identifying enhancement opportunities, enabling community-based suggestion mechanisms.

It is also essential to recognize valuable improvement contributions and foster a sense of shared ownership in the system's direction and evolution.

Organizations with sophisticated feedback and improvement systems that enhance the agent's continuous learning typically achieve better long-term performance and value realization from their digital workforce investments.

Specialization vs. Generalization: Strategic Design Choices

One of the most important design decisions for digital workers is defining their scope of capability, whether to create specialized systems focused on specific domains or more general-purpose agents with broader but potentially shallower capabilities.

Tradeoffs Between Narrow and Broad Capability Sets

The specialization-generalization spectrum presents tradeoffs that must be carefully weighed against organizational needs and implementation context.

Specialization advantages

1. **Performance optimization**

- Higher accuracy within focused domains
- Deeper capability development in specific areas
- More optimized approaches for particular task types
- Greater reliability within scope boundaries

2. **Simplified training and evaluation**

- Clearer success metrics for narrow domains
- More focused training data requirements
- Easier validation of performance
- More straightforward improvement pathways

3. **Reduced complexity**

- Simpler decision spaces with fewer variables
- More predictable behavior patterns
- Clearer boundary conditions and limitations
- Easier integration with specific systems

Generalization advantages

1. Broader applicability

- Ability to handle diverse task types
- Reduced need for multiple specialized systems
- Flexibility to address emerging requirements
- Greater continuity across different processes

2. Contextual understanding

- Ability to connect information across domains
- Recognition of broader patterns and relationships
- Transfer learning across related tasks
- More human-like understanding capabilities

3. Implementation efficiency

- Consolidated development and management
- Reduced integration complexity across systems
- Simplified user experience with fewer interfaces
- Lower overall maintenance requirements

Specialization vs. Generalization

	Specialization	Generalization
Applicability	Higher accuracy, focused domains	Ability to handle diverse tasks
Training & Evaluation	Clearer success metrics	Transfer learning across related tasks
Complexity	Simpler decision spaces	Connect information across domains
Implementation	Easier integration with specific systems	Consolidated development and management
Understanding	More optimized approaches	More human-like understanding capabilities
Scope	Greater reliability within scope	Greater continuity across different processes

Strategic decision factors

Task characteristics play a critical role in shaping digital workforce strategies.

These include the depth versus breadth of knowledge required, the degree of similarity in processing across different use cases, how frequently tasks involve cross-domain connections, and whether task requirements remain stable or evolve over time.

Organizational context looks into available technical expertise and resources, urgency of implementation timelines, and the complexity of the integration environment.

The organization's long-term vision for its digital workforce also contributes to how systems should be designed and deployed.

Technology capabilities must be carefully evaluated as well. Key considerations include the current state of foundation models, availability of specialized systems, integration capabilities across components, and any constraints imposed by development platforms.

Most organizations ultimately develop a portfolio of digital workers with varying degrees of specialization, matching design choices to specific use cases while leveraging common agent components where appropriate.

Domain Adaptation Techniques

Even with a specialized approach, effective digital worker design often involves adapting more general capabilities to specific domains rather than building entirely custom systems from the ground up.

Key domain adaptation approaches

1. **Knowledge enhancement**

- Augmentation with domain-specific information
- Integration of specialized terminology and concepts
- Incorporation of industry-specific rules and constraints
- Connection to authoritative knowledge sources

2. **Task-specific fine-tuning**

- Additional training on domain-relevant examples
- Optimization for particular performance characteristics

- Adjustment of response styles for context
- Specialized evaluation metrics and thresholds

3. **Workflow integration**

- Embedding within domain-specific processes
- Connection to specialized tools and systems
- Adaptation to particular decision sequences
- Alignment with industry-standard approaches

4. **Interface customization**

- Domain-appropriate terminology and conventions
- Specialized input and output formats
- Context-relevant visualization approaches
- User experience aligned with existing tools

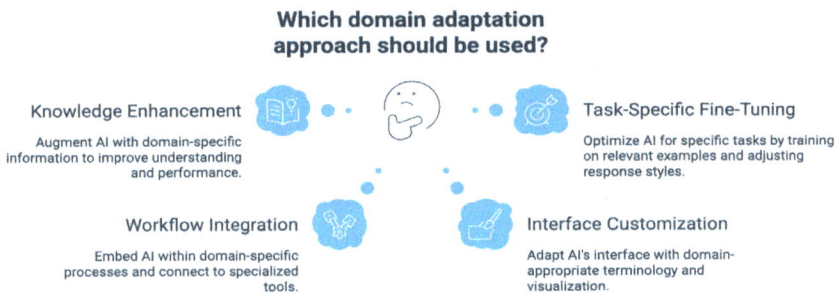

Which domain adaptation approach should be used?

Knowledge Enhancement
Augment AI with domain-specific information to improve understanding and performance.

Task-Specific Fine-Tuning
Optimize AI for specific tasks by training on relevant examples and adjusting response styles.

Workflow Integration
Embed AI within domain-specific processes and connect to specialized tools.

Interface Customization
Adapt AI's interface with domain-appropriate terminology and visualization.

Implementation methodologies

Foundation models: Fine-tuning general-purpose models for specific domains, using prompt engineering to guide domain-specific behavior, augmenting context with specialized knowledge, and formatting outputs to align with domain conventions.

Component composition: Combining general and specialized modules, orchestrating functionality across capability boundaries, dynamically routing tasks based on request characteristics, and progressively specializing general capabilities to meet domain needs.

Hybrid human-AI approaches: Augmenting AI capabilities in specialized areas using progressive automation as domain understanding improves. This approach allows expert review and validation in complex scenarios and fosters collaborative learning between human users and AI systems.

Effective domain adaptation creates specialized capabilities while leveraging the economies of scale and knowledge transfer benefits of more general foundation technologies.

Multi-Capability Orchestration Frameworks

As organizations deploy multiple specialized digital workers, the need for effective orchestration across these systems becomes increasingly important.

Key orchestration requirements

1. **Request routing**

- Appropriate direction of tasks to specialized systems
- Dynamic allocation based on request characteristics
- Load balancing across similar capabilities
- Seamless redirection when initial routing is suboptimal

2. **Context preservation**

- Information sharing across specialized workers
- Consistent user experience across transitions
- Memory persistence between interactions
- Understanding of relationships between different tasks

3. **Composite task handling**

- Decomposition of complex requests into component tasks
- Parallel processing across specialized systems

- Sequential workflows spanning multiple capabilities
- Integration of results into coherent responses

4. **Unified management**

- Consolidated monitoring across diverse workers
- Centralized governance and policy enforcement
- Integrated performance analysis and improvement
- Coordinated evolution and enhancement

Orchestration Requirements Overview

Request Routing

Directing tasks to specialized systems based on characteristics

Context Preservation

Maintaining consistent user experience and information sharing

Composite Task Handling

Managing complex requests through parallel and sequential processing

Unified Management

Centralized monitoring and governance across diverse workers

Orchestration architecture approaches

The central coordinator model involves an intelligent routing layer that operates above specialized workers. This handles request understanding and decomposition, manages task allocation and result integration, and oversees end-to-end process management.

The peer network model facilitates direct communication between specialized systems. It supports capability registration and discovery, enables dynamic collaboration tailored to specific task requirements, and uses distributed coordination protocols.

Hierarchical organization employs a tiered arrangement of generalists and specialists. Requests are routed progressively from general to specific capabilities, with defined escalation pathways for complex cases. This model combines centralized oversight with distributed execution to balance control and scalability.

Orchestration Architecture

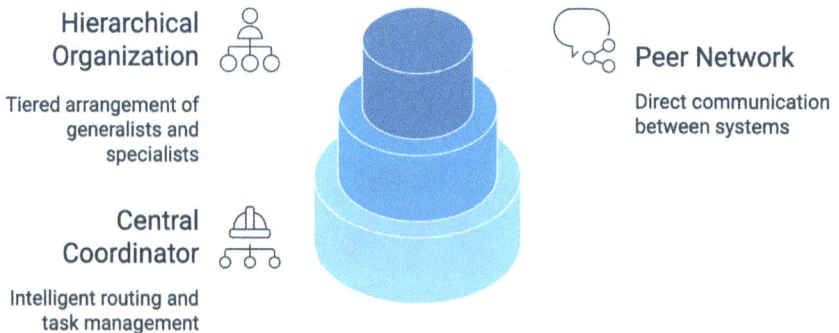

Hierarchical Organization
Tiered arrangement of generalists and specialists

Central Coordinator
Intelligent routing and task management

Peer Network
Direct communication between systems

As digital workforces expand, advanced orchestration frameworks play a critical role in maximizing effectiveness and value.

Conclusion: Designing for Purpose and Evolution

Effective digital worker design is a complex, multidisciplinary challenge that spans **technical architecture, human factors, domain expertise, and strategic alignment**. Organizations that approach this thoughtfully create systems that not only perform immediate tasks effectively but also adapt and evolve as capabilities advance and requirements change.

Several key principles distinguish successful design approaches:

1. **Purpose-driven design**: Start with clear objectives and work backward to determine appropriate capabilities, constraints, and interaction models.

2. **Human-centric approach**: Prioritize collaboration between digital and human workers over simply maximizing automation.

3. **Architectural thinking**: Create modular, extensible systems that evolve over time rather than point solutions to immediate problems.

4. **Balanced specialization**: Define scope thoughtfully based on context and goals, rather than defaulting to either extreme.

5. **Feedback integration**: Build systems that continuously learn and improve based on operational experience and stakeholder input.

As agentic AI capabilities continue to advance, a key differentiator between organizations will be how effectively they design digital workers that align with their specific needs, integrate with their existing systems, and collaborate productively with their human workforce.

CHAPTER 5:

DATA STRATEGIES FOR AGENTIC SYSTEMS

TL;DR:

- Effective digital workers require a comprehensive knowledge foundation that combines different representation approaches to support various types of understanding and reasoning.

- Reliable operation depends on robust data quality management, requiring systematic approaches to input validation, knowledge maintenance, and error detection and correction.

- As digital workers access increasingly sensitive information, robust privacy and security strategies become critical, including appropriate access controls, data minimization techniques, and regulatory compliance approaches.

- Continuous learning capabilities distinguish digital workers from traditional automation, requiring thoughtful strategies for feedback integration, learning method selection, and systematic evolution management.

- The sophistication of underlying data strategies often determines the ultimate success of digital workforce implementations.

A Comprehensive Data Strategy

The effectiveness of agentic AI systems depends on their relationship with data. Unlike traditional rule-based software, digital workers must continuously access, interpret, and learn from diverse information sources.

Data Foundations: Building the Knowledge Base for Digital Workers

Every digital worker requires a knowledge foundation to understand their domain, make appropriate decisions, and act effectively. These foundations must be thoughtfully constructed to support both immediate operational needs and long-term evolution.

Knowledge representation approaches

The way knowledge is represented shapes a digital worker's capabilities, limitations, and potential evolution. Different representation approaches serve different purposes and often complement one another.

Foundational knowledge types

1. **Declarative knowledge**

- Facts, concepts, and relationships about the world
- Domain-specific terminology and definitions
- Organizational policies and procedures
- Contextual information about operating environments

2. **Procedural knowledge**

- Process flows and sequences
- Decision criteria and evaluation methods
- Action patterns for common situations
- Troubleshooting and problem-solving methodologies

3. **Case-Based knowledge**

- Examples of previous situations and resolutions
- Precedents for handling specific scenarios
- Pattern libraries for recognition and adaptation
- Contextual variations and their implications

4. **Normative knowledge**

- Guidelines for appropriate behavior
- Priority frameworks for conflicting considerations
- Ethical boundaries and compliance requirements
- Quality standards and evaluation criteria

Foundations of Knowledge

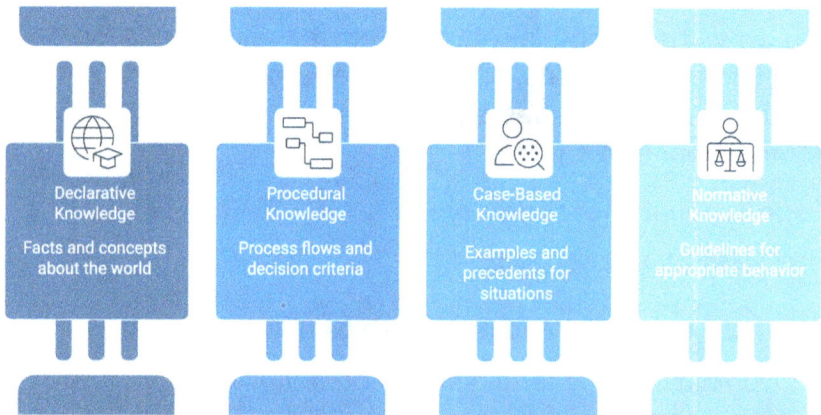

Declarative Knowledge — Facts and concepts about the world

Procedural Knowledge — Process flows and decision criteria

Case-Based Knowledge — Examples and precedents for situations

Normative Knowledge — Guidelines for appropriate behavior

Knowledge representation technologies

1. **Neural representations**

- Distributed knowledge embedded in model parameters
- Implicit relationships discovered through training
- Adaptable patterns that evolve with experience
- Challenging to inspect or directly modify

2. **Symbolic representations**

- Explicit encoding of facts and relationships
- Structured databases of domain knowledge
- Rule systems for procedural understanding
- Transparent and directly editable information

3. **Hybrid approaches**

- Neuro-symbolic integration that combines the strengths of both
- Foundation models augmented with explicit knowledge
- Reasoning systems that leverage implicit and explicit capabilities
- Dynamic balance based on task requirements

Unified Knowledge Systems

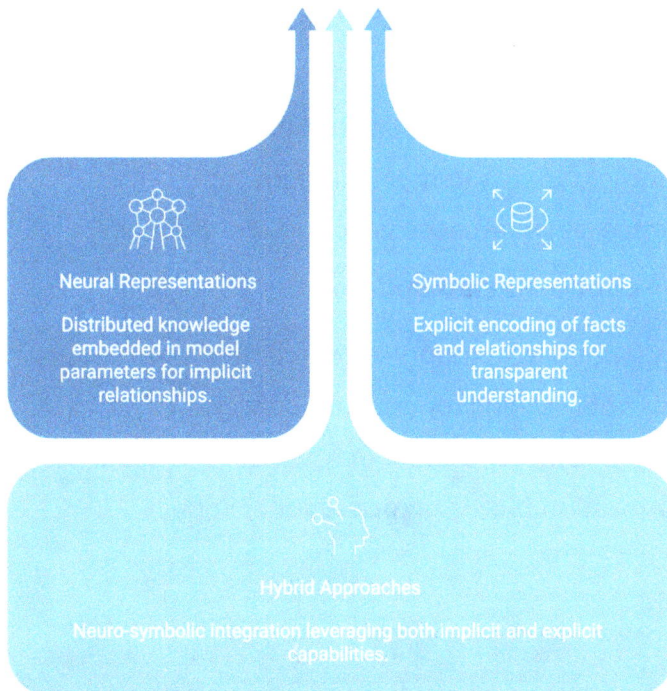

Neural Representations

Distributed knowledge embedded in model parameters for implicit relationships.

Symbolic Representations

Explicit encoding of facts and relationships for transparent understanding.

Hybrid Approaches

Neuro-symbolic integration leveraging both implicit and explicit capabilities.

The most effective knowledge representation strategies combine these approaches, using each where it provides the greatest advantage while compensating for individual limitations.

Information Access and Retrieval Architectures

Agentic systems require architectures that enable them to access, filter, and utilize information from external sources at the point of need.

Access pattern categories

1. **Real-time information retrieval**

- On-demand access to current information
- Search and query across diverse sources
- Contextual filtering based on task requirements
- Dynamic synthesis of retrieved information

2. **Cached knowledge access**

- Local storage of frequently needed information
- Pre-processed data optimized for specific uses
- Periodic updates to maintain currency
- Balance between access speed and freshness

3. **Subscription-based updates**

- Continuous monitoring of key information sources
- Push notifications for relevant changes
- Automatic incorporation of updates
- Prioritization based on operational impact

Retrieval architecture components

1. **Search and indexing systems**

- Efficient location of relevant information
- Semantic understanding of information needs
- Relevance ranking and prioritization
- Multi-source integration and deduplication

2. **Contextual filtering frameworks**

- Task-specific information selection
- Noise reduction and signal enhancement
- Adaptation to user preferences and needs
- Appropriate detail level determination

3. **Knowledge integration mechanisms**

- Synthesis across multiple sources
- Conflict resolution for inconsistent information
- Uncertainty representation in results
- Progressive refinement through interaction

4. **Source selection strategies**

- Authority and reliability assessment
- Recency and relevance evaluation
- Diversity considerations for a broader perspective
- Specialization alignment with specific queries

Advanced retrieval approaches

The evolution of digital workers has coincided with significant advances in retrieval techniques that enable more sophisticated access to knowledge. Understanding these approaches is essential for selecting appropriate architectures for different use cases.

1. **Retrieval-augmented generation (RAG)**

- Combines foundation models with explicit retrieval from external knowledge sources
- Dynamically augments model context with relevant information
- Grounds responses in retrieved facts rather than relying solely on parametric knowledge
- Addresses hallucination risks by providing explicit evidence

Particularly effective for: Dynamic information needs, factual domains requiring high precision, and knowledge-intensive tasks with clear queries.

2. **Retrieval, augmentation, and reasoning engines (RARE)**

- Extends RAG with additional reasoning capabilities
- Performs multi-step decomposition of complex queries
- Orchestrates sequences of retrievals based on intermediate conclusions
- Incorporates explicit reasoning steps to connect retrieved information

Particularly effective for: Complex analytical tasks, multi-hop reasoning requirements, and situations requiring synthesis across domains.

3. **GraphRAG and knowledge graph approaches**

- Leverages structured relationship information rather than just document collections
- Enables navigation across connected entities and concepts
- Supports relationship-aware querying and exploration

- Preserves context through network structure rather than just content similarity

Particularly effective for: Relationship-intensive domains, entity-centric operations, and highly interconnected knowledge domains.

4. **Hierarchical retrieval systems**

- Implements multi-level retrieval processes that begin with broad context and progressively narrow
- Combines coarse document-level and fine passage-level retrieval
- Balances recall and precision through a staged approach
- Often incorporates different retrieval mechanisms at different levels

Particularly effective for: Large-scale knowledge bases, situations requiring both breadth and depth, and complex subject domains.

5. **Hybrid dense-sparse retrieval**

- Combines keyword-based (sparse) and semantic (dense) retrieval approaches
- Leverages the complementary strengths of different retrieval paradigms
- Addresses vocabulary mismatch issues through semantic understanding
- Maintains the precision advantages of lexical matching

Particularly effective for: Technical domains with specialized terminology, situations requiring both precision and recall, and mixed query types.

6. **Multi-agent RERAG (retrieval, augmentation, and reasoning with agentic collaboration)**

- Extends traditional retrieval and reasoning frameworks by coordinating multiple specialized agents
- Enables collaborative problem decomposition, parallel retrieval, and cross-agent reasoning
- Focuses on distinct sub-tasks, such as source selection, fact verification, synthesis, or domain adaptation
- Supports iterative refinement, where agents re-query and augment each other's outputs
- Facilitates advanced information workflows, such as consensus-building, conflict resolution, and scenario-based exploration

Particularly effective for: Highly complex information needs, large-scale research or analysis tasks, and domains requiring diverse expertise or multi-faceted reasoning.

Selection criteria for retrieval approaches

Choosing the appropriate retrieval architecture requires consideration of multiple factors specific to each implementation context:

Knowledge characteristics include considerations such as the size and scale of the knowledge base, the frequency and need for updates, the structure of information (such as text, tables, or relationships), the complexity and interconnectedness of the domain, and the extent of existing organization and metadata.

Task requirements focus on the need to prioritize precision versus recall, any constraints on response time, the complexity of reasoning required, the diversity and variability of queries expected, and the necessity for explanation and supporting evidence.

Technical considerations include the availability of computational resources, acceptable latency and performance requirements, the complexity of integration with existing systems, demands for maintenance and ongoing operations, as well as scalability needs for future growth.

User experience factors include tolerance for retrieval errors, expectations for transparency regarding information sources, the nature of user interaction (whether single queries or conversational dialogue), requirements for personalization, and the typical domain expertise of intended users.

Comprehensive Retrieval Architecture Selection

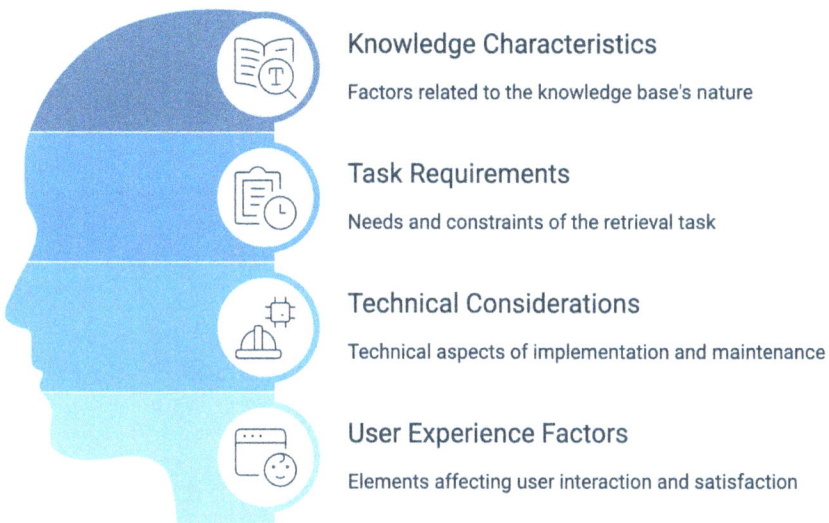

Knowledge Characteristics
Factors related to the knowledge base's nature

Task Requirements
Needs and constraints of the retrieval task

Technical Considerations
Technical aspects of implementation and maintenance

User Experience Factors
Elements affecting user interaction and satisfaction

Effective information access architectures balance performance, accuracy, and maintenance requirements, enabling digital workers to leverage broad knowledge and specialized information appropriate to each situation.

Memory Systems and Context Retention

Unlike traditional software that processes each transaction independently, effective digital workers retain memory across interactions, enabling contextual understanding, relationship building, and continuous improvement.

Memory system categories

1. **Working memory**

- Current conversation or transaction context
- Active goals and priorities
- Immediate task-relevant information
- Short-term considerations and constraints

2. **Episodic memory**

- Historical interactions with specific users
- Previous decisions and their outcomes
- Pattern recognition across similar situations
- Experience-based learning over time

3. **Semantic memory**

- Structured knowledge about domains and topics
- Relationship networks connecting concepts
- Hierarchical categorizations and taxonomies
- Abstract principles and generalizations

4. **Procedural memory**

- Learned processes and sequences
- Refined approaches based on experience
- Optimized patterns for common situations
- Adaptive techniques for handling variations

Memory Systems

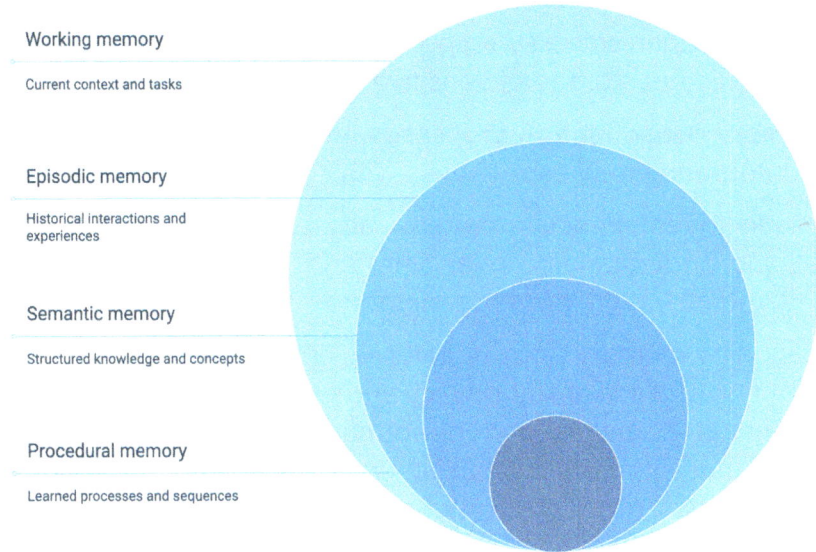

Working memory

Current context and tasks

Episodic memory

Historical interactions and
experiences

Semantic memory

Structured knowledge and concepts

Procedural memory

Learned processes and sequences

Memory implementation approaches

Explicit storage systems use dedicated databases to record interaction histories and store structured records of past decisions and outcomes.

They also include searchable repositories of previous cases and knowledge management systems designed to support the reuse of established patterns and solutions across similar scenarios.

Embedding and retrieval models offer a more dynamic memory mechanism by using vector-based representations of interactions and domain knowledge.

These models enable similarity-based retrieval of past experiences, support contextual filtering for situation-appropriate memory, and apply dynamic weighting to prioritize information based on both recency and relevance.

Forgetting mechanisms involve prioritizing the retention of high-value information, allowing less relevant data to decay over time, and abstracting specific instances into general patterns that can guide future behavior.

These mechanisms also respect ethical boundaries by limiting the persistent storage of sensitive information.

Memory consolidation processes help digital workers evolve by integrating new experiences with existing knowledge.

This includes extracting patterns across multiple interactions, refining understanding over time, and periodically reviewing and enhancing the system's memory.

Designing memory systems involves critical tradeoffs between retention and scalability, privacy and personalization, and computational demands and performance. Organizations must weigh these thoughtfully based on their specific use cases and constraints.

Data Quality Management: Ensuring Reliable Operation

The reliability of digital workers depends on the quality of the data they access, process, and learn from. Organizations must implement comprehensive data quality management strategies to ensure these systems operate effectively.

Input data validation mechanisms

Digital workers, particularly those based on foundation models, are sensitive to input quality issues. Robust validation mechanisms are essential to identify and address problematic inputs.

Input validation approaches

1. **Structural validation**

- Format and schema conformance
- Completeness assessment for required fields
- Type and range verification for values

- Consistency across related inputs

2. **Content quality assessment**

- Relevance evaluation for the intended purpose
- Language quality and comprehensibility
- Bias and problematic content detection
- Factual accuracy verification, where possible

3. **Source reliability evaluation**

- Authority assessment of information sources
- Consistency across multiple sources
- Historical reliability tracking of sources
- Confidence scoring based on source characteristics

Input Validation Approaches

Structural validation

Format and schema conformance, completeness assessment, type and range verification, consistency across related inputs.

Relevance evaluation, language quality and comprehensibility, bias and problematic content detection, factual accuracy verification.

Content quality assessment

Source reliability evaluation

Authority assessment, consistency across sources, historical reliability tracking, confidence scoring based on source characteristics.

Implementation mechanisms

1. **Pre-processing filters**

- Automated quality checks before processing
- Standardization and normalization procedures
- Error correction for common issues
- Enrichment with missing information

2. **Interactive validation**

- Clarification requests for ambiguous inputs
- Confidence-based processing decisions
- User confirmation for critical information
- Progressive refinement through dialogue

3. **Multi-stage processing**

- Initial analysis to detect potential issues
- Specialized handling for problematic cases
- Alternative processing paths based on quality
- Escalation protocols for severely degraded inputs

Implementing robust input validation leads to greater reliability, more consistent performance, and reduced risk of problematic outputs from their digital workers.

Knowledge Base Maintenance Strategies

The knowledge foundations of digital workers require ongoing maintenance to remain accurate, complete, and current. Effective maintenance strategies balance comprehensiveness with operational efficiency.

Key maintenance requirements

1. **Currency management**

- Identification of time-sensitive information
- Update frequency determination based on volatility
- Change detection mechanisms for critical knowledge
- Version control across knowledge components

2. **Consistency assurance**

- Conflict detection across knowledge sources
- Relationship integrity maintenance
- Terminology standardization and alignment
- Coherence checking across domains

3. **Coverage expansion**

- Gap identification in existing knowledge
- Prioritization of enhancement opportunities
- Systematic expansion in high-value areas
- Balance between breadth and depth

4. **Relevance optimization**

- Usage pattern analysis to identify valuable knowledge
- Retirement of obsolete or unutilized information
- Emphasis adjustment based on operational importance
- Alignment with evolving business priorities

Knowledge Maintenance Strategies

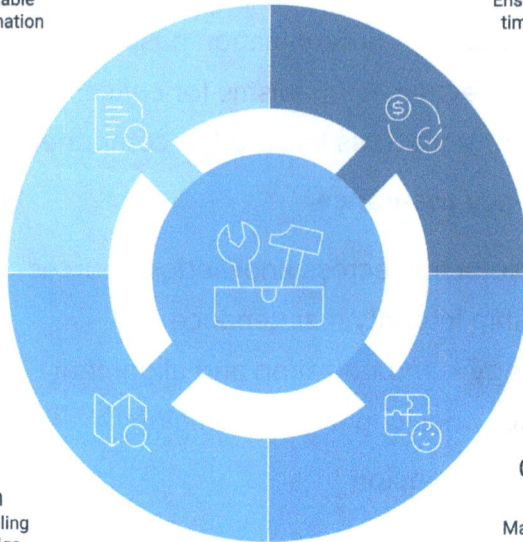

Relevance Optimization
Focusing on valuable and current information

Currency Management
Ensuring information is timely and up-to-date

Coverage Expansion
Identifying and filling gaps in knowledge

Consistency Assurance
Maintaining integrity and coherence across knowledge sources

Maintenance process approaches

Scheduled review cycles involve regular, comprehensive assessments of the knowledge base to ensure accuracy and relevance. These cycles include structured update processes for different knowledge types, periodic validation against authoritative sources, and systematic documentation of all changes made.

Continuous monitoring involves automated detection of potential issues and real-time updates for critical information. It also supports performance-based identification of knowledge gaps and integrates user feedback to continuously refine and improve the knowledge base.

Expert curation relies on domain specialists to oversee specific knowledge areas. This approach includes structured review and approval processes, authority-based resolution of conflicts, and rigorous quality assessments before new information is incorporated into the system.

The most effective knowledge maintenance combines these approaches, tailoring strategies to each knowledge component based on their criticality, volatility, and impact on performance.

Error Detection and Correction Methodologies

Despite the best preventive measures, errors are inevitable in complex information environments. Effective digital worker implementations include robust approaches for detecting and correcting these errors.

Error categories and detection methods

1. **Factual errors**

- Consistency checking against authoritative sources
- Cross-validation across multiple information points
- Anomaly detection for unexpected assertions
- Confidence scoring for factual claims

2. **Reasoning failures**

- Logical consistency validation
- Path analysis for decision processes
- Benchmark comparison for standard problems
- Performance pattern monitoring for degradation

3. **Execution mistakes**

- Outcome validation against expectations
- Process adherence verification
- Result quality assessment
- Deviation detection from historical patterns

Error Categories and Detection Methods

Factual errors	Reasoning failures	Execution mistakes
Consistency checking against authoritative sources, cross-validation across multiple information points, anomaly detection for unexpected assertions, confidence scoring for factual claims.	Logical consistency validation, path analysis for decision processes, benchmark comparison for standard problems, performance pattern monitoring for degradation.	Outcome validation against expectations, process adherence verification, result quality assessment, deviation detection from historical patterns.
1	2	3

Correction approaches

Immediate remediation focuses on addressing errors in real time as they occur during processing. This includes selecting alternative paths when errors are detected, applying graceful degradation strategies instead of complete failure, and maintaining transparent communication about limitations in the system's output.

Learning-based improvement includes recognizing recurring error patterns, conducting root cause analysis to identify systematic issues, adjusting models based on those patterns, and implementing targeted enhancements in problem-prone areas to prevent recurrence.

Human collaboration includes expert review of complex error cases, integrating user feedback to correct mistakes, applying supervised learning from human interventions, and establishing escalation protocols for critical issues that require immediate human oversight or decision-making.

Organizations that implement strong error management achieve greater resilience, faster performance improvement, and higher user trust in their digital workforce implementations.

Privacy and Security: Protecting Sensitive Information

As digital workers access increasingly sensitive information, ensuring privacy and security becomes a critical aspect of data strategy. Effective approaches balance utility with protection, enabling valuable functionality while managing risks.

Access Control Frameworks for Digital Workers

Digital workers require carefully designed access control systems that account for their unique operational characteristics and associated risks.

Access control design principles

1. **Least privilege orientation**

- Minimal access provision for required functions
- Capability-specific permission scoping
- Temporary elevation for specific tasks
- Regular review and adjustment of privileges

2. **Contextual authorization**

- Task-specific access determination
- User relationship considerations
- Time and location factors
- Purpose limitation enforcement

3. Granular control

- Field-level rather than just resource-level permissions
- Operation-specific authorizations
- Data category-based restrictions
- User-driven permission management

Access Control Design Principles

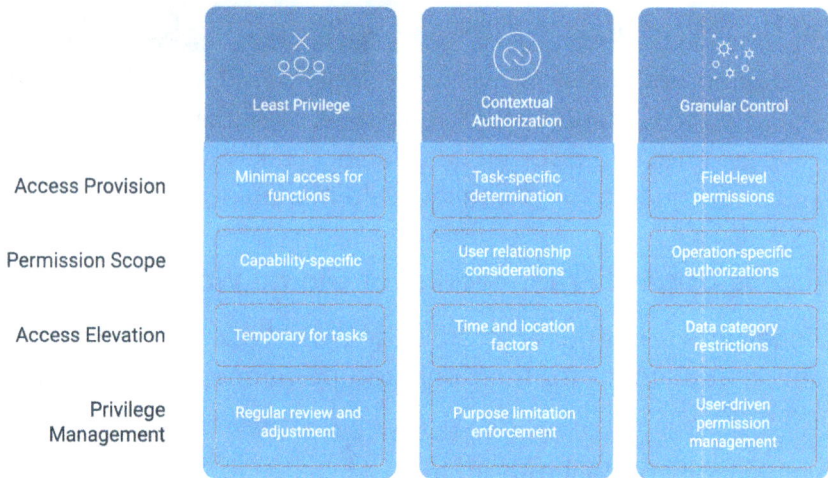

	Least Privilege	Contextual Authorization	Granular Control
Access Provision	Minimal access for functions	Task-specific determination	Field-level permissions
Permission Scope	Capability-specific	User relationship considerations	Operation-specific authorizations
Access Elevation	Temporary for tasks	Time and location factors	Data category restrictions
Privilege Management	Regular review and adjustment	Purpose limitation enforcement	User-driven permission management

Implementation approaches

Identity and authentication systems include establishing secure digital worker identity, enforcing appropriate authentication mechanisms, managing and rotating credentials to mitigate risk, and implementing session control and timeout policies to prevent unauthorized access during idle periods.

Authorization frameworks incorporate role-based access control and attribute-based permission models, enforce policies at various points within the workflow, and maintain centralized governance with distributed enforcement.

Delegation mechanisms enable users to grant access directly to digital workers, provide temporary permissions, and offer transparent visibility into the access scope.

Organizations must recognize that digital worker access control requires different approaches than traditional human or system access management.

Data Minimization Techniques

An important principle of responsible digital worker implementation is data minimization, ensuring these systems access and retain only the information necessary for their intended functions.

Minimization strategy components

1. **Purpose specification**

- Clear articulation of data usage purposes
- Limitation to specifically defined functions
- Prevention of scope creep and function expansion
- Regular validation of continuing necessity

2. **Collection limitation**

- Selective gathering of required information
- Alternatives assessment for less sensitive data
- Progressive collection based on actual needs
- User control over optional information sharing

3. **Processing restrictions**

- Function-specific data utilization
- Prevention of incidental learning from sensitive data
- Separation between operational and training usage
- Prohibition of unauthorized data exploitation

4. **Retention control**

- Time-limited storage of transaction data
- Scheduled purging of unnecessary information
- Anonymization after immediate needs fulfilled

- Justification requirements for extended retention

Data Minimization Process

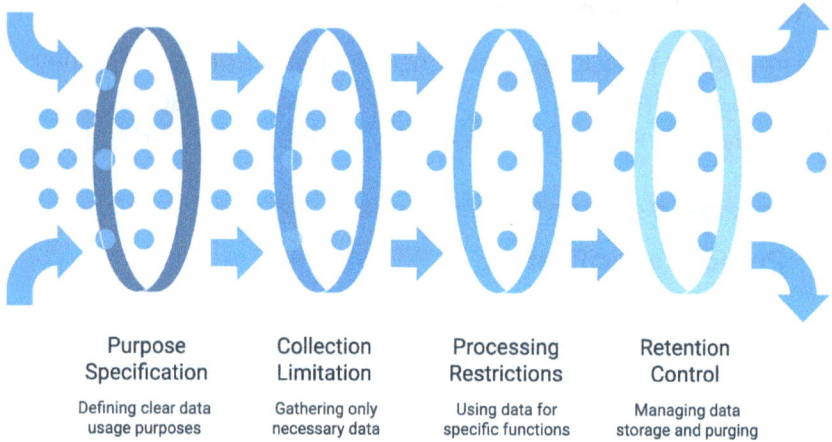

Purpose Specification	Collection Limitation	Processing Restrictions	Retention Control
Defining clear data usage purposes	Gathering only necessary data	Using data for specific functions	Managing data storage and purging

Implementation mechanisms

1. **Data filtering systems**

- Pre-processing to remove unnecessary information
- Real-time redaction of sensitive content
- Context-appropriate detail level determination
- Selective masking based on function requirements

2. **Differential privacy approaches**

- Noise addition to prevent individual identification
- Aggregation techniques for statistical usage
- Query limitation to protect individual privacy
- Formal privacy guarantees for sensitive applications

3. **Federated processing models**

- Local computation on sensitive data when possible
- Minimal information transfer across boundaries

- Result abstraction rather than raw data sharing
- Decentralized learning without central data collection

Effective data minimization not only reduces privacy and security risks but also improves performance by focusing on truly relevant information.

Compliance Strategies for Regulatory Requirements

Digital workers operate within increasingly complex regulatory environments, making comprehensive compliance strategies essential to navigate these requirements.

Key regulatory domains

1. **Privacy regulations**

- Personal data protection requirements
- Consent and transparency obligations
- Cross-border transfer restrictions
- Individual rights fulfillment processes

2. **Industry-specific requirements**

- Financial services regulations
- Healthcare information protection
- Critical infrastructure security standards
- Professional services confidentiality rules

3. **Algorithmic governance**

- Fairness and non-discrimination requirements
- Explainability and transparency obligations
- Impact assessment mandates
- Accountability and oversight mechanisms

Regulatory Domains

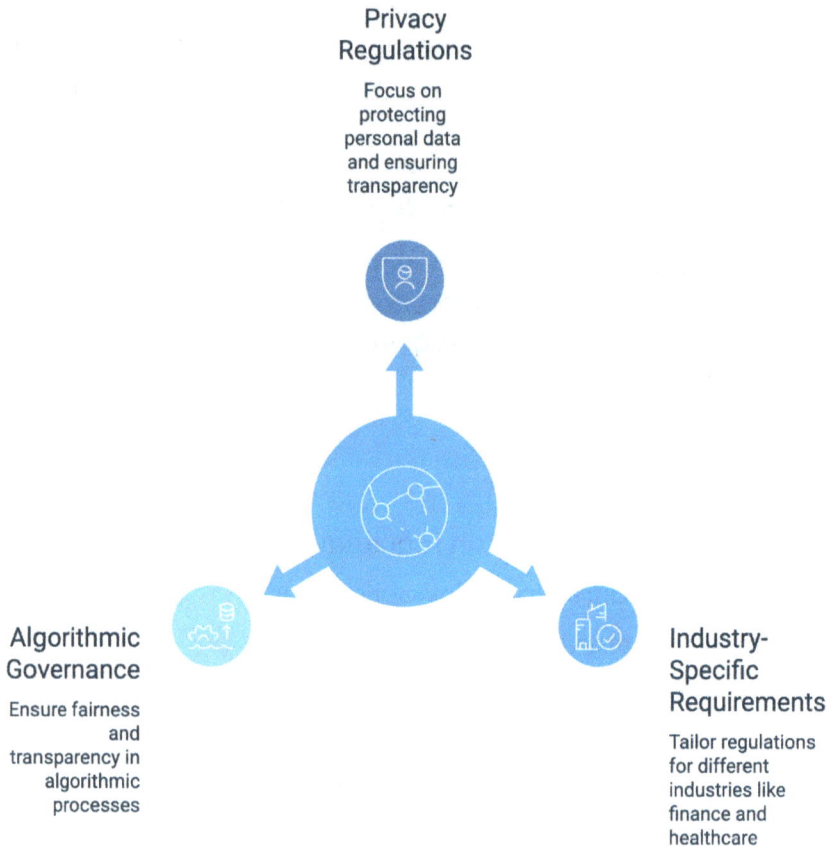

Privacy Regulations

Focus on protecting personal data and ensuring transparency

Algorithmic Governance

Ensure fairness and transparency in algorithmic processes

Industry-Specific Requirements

Tailor regulations for different industries like finance and healthcare

Compliance implementation approaches

Privacy by design embeds compliance into the system's initial architecture, ensuring that privacy protections are active by default, without requiring user intervention.

It also includes systematic impact assessments during the development phase and ongoing compliance validation throughout the digital worker's lifecycle.

Automated compliance controls enable policy enforcement through technical mechanisms.

These controls include real-time monitoring to detect potential violations, preventive restrictions on problematic actions, and comprehensive audit logging to support verification and demonstration.

Governance framework integration involves assigning clear accountability for compliance responsibilities, establishing regular review processes to ensure ongoing alignment with regulations, and maintaining thorough documentation of design decisions and control implementations.

To proactively manage compliance-related threats, systematic risk assessment and mitigation strategies are also employed.

Organizations must recognize that regulatory requirements for digital workers are evolving, demanding adaptive compliance strategies that anticipate emerging obligations.

Continuous Learning: Approaches for Ongoing Improvement

Unlike traditional automation, digital workers continuously learn and improve. Effective implementations include comprehensive strategies to leverage this capability while managing associated risks.

Feedback Loop Design for Performance Enhancement

Systematic feedback integration enables digital workers to improve based on operational experience, gradually enhancing their effectiveness and reliability.

Feedback source categories

1. **Direct user feedback**

- Explicit evaluations of responses or actions
- Correction of errors or misunderstandings
- Preference expressions for future interactions

- Suggestions for improvement opportunities

2. **Implicit behavioral signals**

- Usage patterns indicating satisfaction or frustration
- Selection among alternative options
- Engagement depth with the provided information
- Repeat usage patterns for similar situations

3. **Outcome measurement**

- Goal achievement assessment
- Efficiency metrics for processes
- Quality evaluation of generated outputs
- Business impact measurement for actions

4. **Expert review**

- Specialized evaluation of complex outputs
- Detailed assessment of reasoning processes
- Comparative analysis against best practices
- Identification of subtle improvement opportunities

Feedback sources range from direct to indirect input.

Direct `<`   `>` Indirect

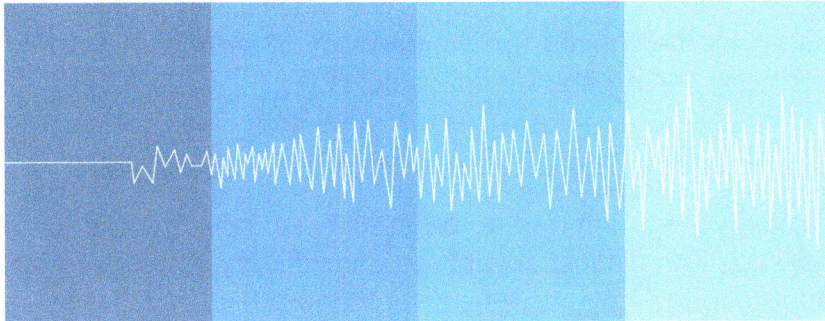

Direct User Feedback	Implicit Behavioral Signals	Outcome Measurement	Expert Review
Explicit evaluations and corrections	Usage patterns indicate satisfaction	Goal achievement and efficiency metrics	Specialized evaluation and comparative analysis

Feedback integration mechanisms

1. **Real-time adaptation**

- Immediate adjustment based on interaction
- Dynamic response refinement within sessions
- Progressive personalization for specific users
- Context-specific behavior modification

2. **Periodic refinement cycles**

- Aggregated feedback analysis across interactions
- Pattern identification for systematic issues
- Prioritized enhancement based on impact
- Scheduled capability updates and improvements

3. **Supervised learning integration**

- Expert-guided improvement for complex capabilities

- Specialized training for challenging domains
- Targeted enhancement for identified weaknesses
- Continuous capability expansion in high-value areas

Effective feedback integration combines these approaches based on the specific requirements of different digital worker functions and organizational contexts.

Supervised vs. Unsupervised Learning Strategies

Digital workers improve through different learning approaches, each with distinct advantages, limitations, and appropriate applications.

Learning approach characteristics

1. **Supervised learning**

- Relies on explicitly labeled examples
- Provides clear guidance for desired behavior
- Enables precise control over capability development
- Requires significant human involvement and expertise
- Particularly valuable for critical functions with clear, correct answers

2. **Reinforcement learning**

- Based on reward signals for desired outcomes
- Enables optimization toward specific objectives
- Allows discovery of novel approaches
- Requires careful reward function design
- Valuable for tasks with clear success metrics but multiple potential approaches

3. **Unsupervised learning**

- Identifies patterns without explicit guidance

- Discovers relationships and structures autonomously
- Enables insight generation beyond predetermined categories
- Provides limited control over specific outcomes
- Useful for exploration and knowledge discovery in complex domains

Which learning approach should be used for the AI project?

Unsupervised Learning

Suitable for exploration and knowledge discovery in complex domains, with limited control.

Reinforcement Learning

Ideal for tasks with clear success metrics, allowing optimization and discovery.

Supervised Learning

Best for tasks requiring precise control and clear guidance, but needs significant human involvement.

Strategic application approaches

Hybrid learning models combine approaches based on task characteristics. These models begin with a supervised foundation of reinforcement learning to refine behaviors. In some cases, unsupervised exploration is allowed within supervised boundaries.

As the digital worker's capabilities mature, these models support progressive autonomy, allowing for more independent operation over time.

Risk-based learning selection emphasizes more controlled approaches for high-consequence functions, while low-risk or exploratory areas can tolerate greater autonomy and experimentation.

Oversight is applied in layesrs, proportional to the potential impact of errors, and may be gradually relaxed as the system consistently demonstrates reliability and performance.

Domain-specific optimization ensures that learning strategies are customized based on the unique needs of each domain.

This involves selecting learning methods that align with the available data and subject-matter expertise, incorporating requirements for explainability where necessary, and adapting to regulatory or compliance constraints that may influence how knowledge is acquired and applied.

Organizations must develop nuanced learning strategies that reflect their unique contexts, capabilities, and risk profiles rather than applying universal approaches across all functions.

Model Refresh and Update Methodologies

As digital workers continuously learn and evolve, organizations must implement systematic approaches to manage this evolution, ensuring stability, reliability, and alignment with objectives.

Refresh requirement categories

1. **Performance improvement**

- Capability enhancement based on operational feedback
- Error pattern remediation through targeted updates
- Efficiency optimization for frequently used functions
- Extension to handle edge cases and exceptions

2. **Knowledge currency**

- Incorporation of updated information
- Alignment with evolving best practices
- Addition of new subject domains
- Removal of obsolete or incorrect information

3. **Alignment maintenance**

- Reinforcement of value consistency
- Correction of developing biases or skews
- Adaptation to evolving organizational priorities
- Response to changing regulatory requirements

Requirement Categories

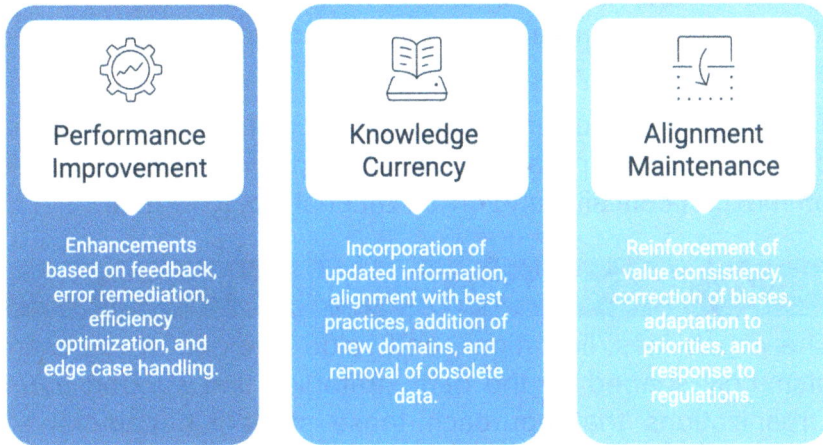

Performance Improvement	Knowledge Currency	Alignment Maintenance
Enhancements based on feedback, error remediation, efficiency optimization, and edge case handling.	Incorporation of updated information, alignment with best practices, addition of new domains, and removal of obsolete data.	Reinforcement of value consistency, correction of biases, adaptation to priorities, and response to regulations.

Update implementation approaches

Continuous versus periodic updates balance adaptability with stability. Real-time learning is valuable for rapidly evolving functions that benefit from immediate adaptation. In contrast, stability-critical applications rely on scheduled updates to maintain predictability and control.

Hybrid approaches are also common, where different components follow different update cycles based on their roles. Additionally, emergency update protocols are essential for quickly addressing critical issues.

Staged deployment strategies help ensure smooth rollouts. Updates are progressively released across environments, starting with limited exposure through canary testing to detect issues early.

A/B testing is used to compare the performance with fallback mechanisms to revert to stable configurations if unexpected issues arise.

Version control and governance include comprehensive tracking of how models evolve over time, with clear authorization processes for implementing changes.

Documentation helps capture the rationale, expected impacts, and performance metrics for each version to support informed decision-making.

The most effective refresh strategies balance continuous improvement with operational stability, applying controls based on function criticality and organizational risk tolerance.

Conclusion: Data as the Foundation of Agentic Value

The capabilities, limitations, and value of digital workers are ultimately shaped by the data strategies that support them. Organizations that approach these strategies comprehensively, addressing knowledge representation, quality management, privacy protection, and continuous learning, create digital workforces that not only perform effectively today but continuously evolve to deliver greater value over time.

Several key principles distinguish successful approaches:

1. **Strategic integration**: Aligning data strategies with broader digital workforce objectives and organizational priorities rather than treating them as isolated technical considerations.

2. **Balanced protection and utility**: Enabling valuable functionality while providing appropriate safeguards for sensitive information and compliance requirements.

3. **Dynamic evolution**: Creating systems that continuously learn and improve through structured feedback loops while maintaining stability and reliability.

4. **Ethical foundation**: Prioritizing privacy, fairness, and appropriate human oversight and control.

5. **Architectural thinking**: Designing data foundations that enable current needs while providing flexibility for future evolution and expansion.

As digital workforces mature, the sophistication of underlying data strategies will increasingly differentiate organizations that capture their full potential from those that achieve only marginal benefits.

Investing in these foundations, while less visible than the digital workers themselves, often determines the ultimate success of implementation initiatives.

CHAPTER 6:
DEPLOYMENT AND SCALING

TL;DR:

- Successful implementation typically begins with carefully designed pilots that demonstrate value while establishing foundations for broader deployment.

- Comprehensive change management is essential, addressing both emotional responses and practical adaptation needs through strategic communication, training, and resistance management.

- Scaling beyond initial pilots requires thoughtful approaches to both technical architecture and organizational governance to enable coordinated expansion while maintaining quality and consistency.

- Ensuring sustained performance demands sophisticated monitoring, issue management, and continuous improvement processes that enable digital workers to evolve rather than stagnate.

- Organizations that excel at deployment and scaling recognize implementation as a transformation journey requiring integrated evolution of technology, processes, skills, and mindsets.

The Path to a Scalable Digital Workforce

Designing effective digital workers is only part of the implementation journey. The true test comes during deployment, as systems move from controlled development environments into the complex reality of organizational operations.

Pilot Implementation: Starting Small and Proving Value

Deployment typically begins with carefully designed pilot implementations that demonstrate value, refine strategies, build organizational confidence, and establish the foundation for broader expansion.

Scope Definition for Initial Deployments

Effective pilot implementations begin with a well-defined scope that balances ambitious goals to demonstrate meaningful value with sufficient constraints to enable success. Key scoping considerations include:

1. **Process selection criteria**

- Meaningful but manageable complexity
- Sufficient volume to demonstrate value
- Relatively stable and well-documented procedures
- Moderate but not extreme business criticality
- Reasonable stakeholder alignment and support

2. **Functional boundary definition**

- Clear delineation of included versus excluded capabilities
- Specific user populations for initial access
- Explicit handling procedures for edge cases
- Defined interfaces with adjacent systems
- Phasing approach for feature introduction

3. **Establish a success threshold**

- Minimum viable functionality requirements
- Performance expectations for progression
- Key metrics for value demonstration
- Timeframes for assessment and decision-making
- Comparison baselines for improvement measurement

4. **Resource allocation parameters**

- Implementation team composition and time commitment
- Stakeholder involvement expectations
- Technical infrastructure requirements
- Timeline and milestone definitions
- Budget constraints and tracking mechanisms

Steps to Effective Pilot Implementation

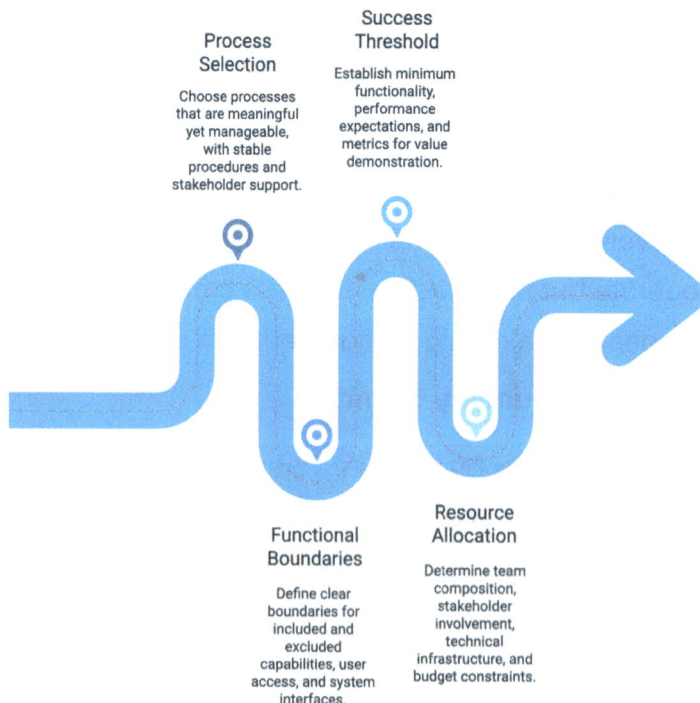

Process Selection

Choose processes that are meaningful yet manageable, with stable procedures and stakeholder support.

Success Threshold

Establish minimum functionality, performance expectations, and metrics for value demonstration.

Functional Boundaries

Define clear boundaries for included and excluded capabilities, user access, and system interfaces.

Resource Allocation

Determine team composition, stakeholder involvement, technical infrastructure, and budget constraints.

Controlled Environment Testing

Before deployment, effective implementations undergo rigorous testing in controlled environments that progressively simulate real-world conditions. This includes:

1. **Development environment testing**

- Basic functionality validation
- Component-level performance assessment
- Integration point verification
- Security and compliance control validation
- Developer-driven exploratory testing

2. **Simulated environment testing**

- End-to-end process validation
- Performance assessment under expected load
- Data variety and edge case handling
- Integration with replicated connected systems
- Controlled exception introduction and handling

3. **Staging environment validation**

- Testing with production-equivalent infrastructure
- Integration with actual (but duplicated) systems
- Real data volume and variety exposure
- Full security and compliance control activation
- Operational procedure verification

4. **Limited production testing**

- Controlled exposure to actual business operations
- Parallel processing alongside existing approaches
- Gradual expansion of the user population

- Progressive removal of safety constraints
- Continuous monitoring and rapid response capabilities

Each testing stage should include specific objectives, success criteria, and explicit decision gates for progression to subsequent phases. Testing should include not only technical functionality but also user experience, operational procedures, and governance mechanisms.

Success Metrics and Evaluation Frameworks

Effective pilot implementations use comprehensive evaluation frameworks that measure multiple dimensions of performance.

Key performance dimensions

1. **Technical performance assessment**

- Accuracy and quality of outputs
- Processing time and throughput capacity
- Exception frequency and handling effectiveness
- System stability and reliability
- Security and compliance control effectiveness

2. **Business impact measurement**

- Process efficiency improvements
- Cost reduction realization
- Quality and consistency enhancements
- Capacity and throughput increases
- Revenue enhancement or protection

3. **User experience evaluation**

- Adoption and utilization patterns
- User satisfaction measurements

- Training effectiveness assessment
- Support requirement monitoring
- Feedback theme analysis

4. **Organizational readiness indicators**

- Skills and capability development progress
- Process and procedure effectiveness
- Governance mechanism validation
- Stakeholder engagement and support evolution
- Cultural acceptance and alignment

Framework for Performance Evaluation

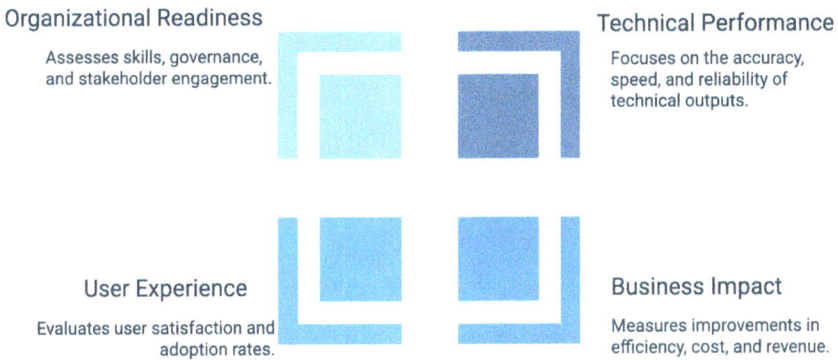

Organizational Readiness
Assesses skills, governance, and stakeholder engagement.

Technical Performance
Focuses on the accuracy, speed, and reliability of technical outputs.

User Experience
Evaluates user satisfaction and adoption rates.

Business Impact
Measures improvements in efficiency, cost, and revenue.

Measurement approaches

Baseline comparison methodologies include measuring outcomes before and after implementation, using control groups when feasible to isolate the impact, and analyzing historical trends for contextual understanding.

Comparing expected outcomes against actual results helps identify deviations, while industry benchmarks, when available, offer external standards to gauge relative performance.

Mixed-method evaluation incorporates quantitative performance metrics alongside qualitative user and stakeholder feedback.

It also includes structured observation of operational patterns and process mining or interaction analysis to uncover underlying trends, while indirect impact assessment looks at how implementations affect adjacent processes.

Combining multiple approaches creates a holistic understanding of pilot performance rather than focusing on isolated metrics that may present a misleading picture of success or failure.

Stakeholder Management During Early Stages

The technical success of pilot implementations often depends as much on stakeholder management as on the underlying technology. Early deployments require careful effort to build and maintain support across diverse organizational groups.

Key stakeholder categories

1. **Executive sponsors**

- Providing organizational legitimacy and support
- Securing necessary resources and removing barriers
- Communicating strategic alignment and importance
- Setting appropriate expectations with peer leaders
- Maintaining commitment through implementation challenges

2. **Front-line users**

- Directly engaging with digital worker capabilities
- Providing essential feedback on functionality and usability
- Adapting daily workflows and procedures
- Serving as peers to influence other potential users
- Experiencing immediate impact on daily work experience

3. **Technical and support teams**

- Maintaining system functionality and performance

- Providing user assistance and issue resolution
- Implementing enhancements and refinements
- Managing integration with connected systems
- Developing expertise for subsequent deployment phases

4. **Process owners and middle management**

- Aligning departmental priorities and resource allocation
- Adapting performance expectations and metrics
- Managing workflow transitions during implementation
- Supporting team members through change
- Validating business value realization

Stakeholder Roles and Responsibilities

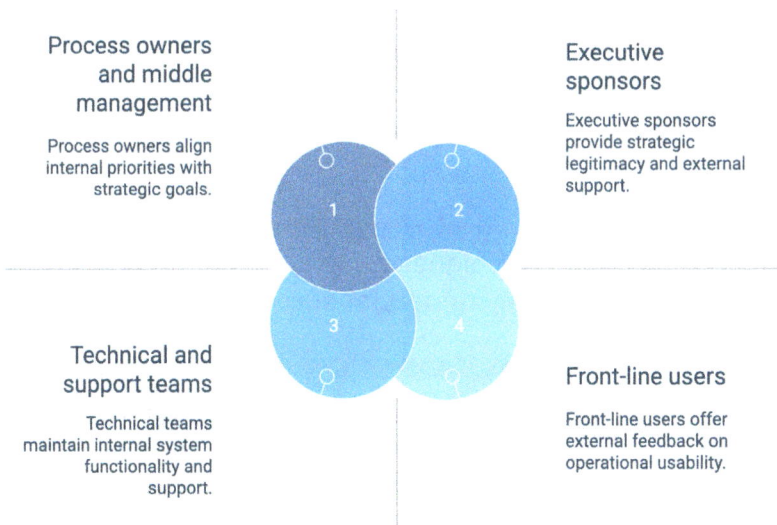

Process owners and middle management

Process owners align internal priorities with strategic goals.

Executive sponsors

Executive sponsors provide strategic legitimacy and external support.

Technical and support teams

Technical teams maintain internal system functionality and support.

Front-line users

Front-line users offer external feedback on operational usability.

Engagement strategies

A transparent communication methodology is key to building trust. This includes clearly articulating the purpose and expected benefits, honestly discussing potential challenges and limitations, and providing regular updates on progress and emerging insights.

It also means explicitly acknowledging concerns and questions, and fostering a two-way dialogue rather than relying solely on top-down announcements.

Involvement and co-creation strengthen engagement by enabling early participation in design and planning. Stakeholders should have meaningful input into key decisions, with their expertise recognized and applied.

Encouraging collaborative problem-solving for emerging issues and promoting shared ownership of the implementation supports long-term success.

Managing expectations involves realistic framing of capabilities and limitations, clear communication about implementation timelines, and transparent discussions of any potential disruptions.

Gradually disclosing increasing capabilities while maintaining a balanced view of benefits and challenges helps build confidence and maintain alignment.

Organizations that excel at stakeholder management during pilots typically achieve higher adoption rates, more valuable feedback, and stronger support for subsequent scaling efforts.

Change Management: Facilitating Organizational Adoption

The introduction of digital workers is a significant change to established organizational patterns. Successful implementation requires comprehensive change management approaches that address both emotional and practical dimensions of adaptation.

Communication Strategies for Digital Workforce Adoption

Strategic communication is central to change management, shaping perceptions, expectations, and understanding across the organization.

Communication framework elements

Narrative development involves a compelling explanation of the "why" behind the implementation.

It should connect the initiative to broader organizational strategy and objectives, balance opportunity and necessity, and acknowledge legitimate concerns alongside benefits. Personalization of messaging for different stakeholder perspectives ensures greater relevance and engagement.

Multi-channel distribution includes formal organizational communications, leadership messaging in established forums, team-level discussions, and one-on-one conversations with key stakeholders. Demonstrations and experiential opportunities further reinforce understanding.

Messaging progression should follow a logical flow, starting with building initial awareness and setting context, followed by deeper understanding development through examples. It then moves into clarifying specific impacts and expectations, preparing stakeholders for change, and reinforcing success stories.

Feedback integration includes active listening, collecting and transparently addressing questions, identifying and correcting misconceptions, and continuously adapting messages based on stakeholder response and evolving understanding.

Key message themes

Framing augmentation versus replacement involves emphasizing how digital tools enhance human capabilities rather than replace them. It focuses on eliminating low-value tasks, highlights examples of new opportunities created, and reinforces the importance of continued human judgment. Concrete illustrations of human-digital collaboration help bring this message to life.

Transition support assurance centers on communicating the assistance available during change. This includes outlining specific training and development opportunities, acknowledging the challenges of transition, and providing clear access to support resources. Leadership commitment plays a key role in reassuring teams of sustained support.

Emphasizing employee involvement highlights the importance of giving people a voice in implementation. It underscores opportunities to shape implementation, values front-line expertise, and encourages feedback and experience sharing. The framing promotes co-creation and fosters a sense of collective ownership over the change.

Navigating Digital Transformation with Key Messages

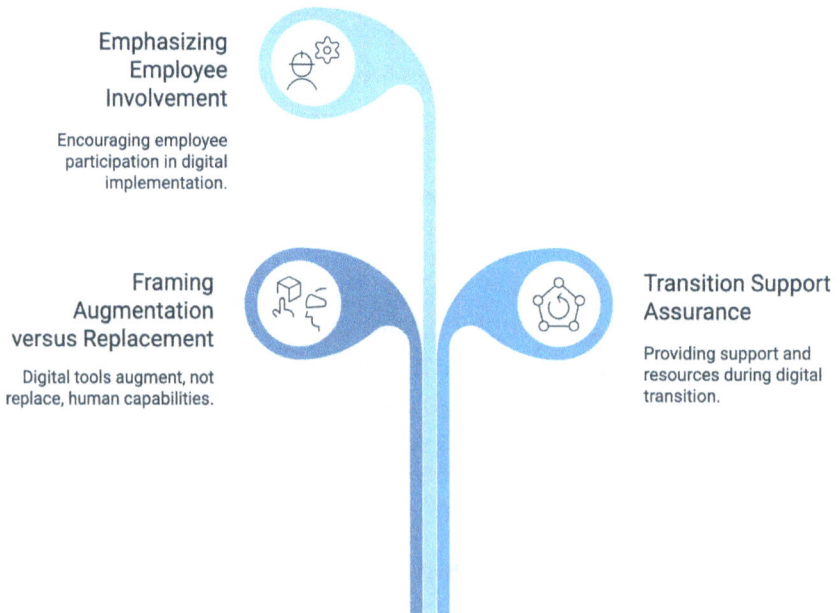

Emphasizing
Employee
Involvement

Encouraging employee
participation in digital
implementation.

Framing
Augmentation
versus Replacement

Digital tools augment, not
replace, human capabilities.

Transition Support
Assurance

Providing support and
resources during digital
transition.

Effective communication strategies recognize that digital work-force adoption isn't merely a technical change but a profound shift in how work happens across the organization.

Training Approaches for Human Collaborators

The effectiveness of digital workers depends significantly on the ability of their human colleagues to collaborate productive-ly with them. Comprehensive training strategies address both technical skills and mindset evolution.

Training component categories

1. **Understanding development**

- Basic digital worker capabilities and limitations
- Key terminology and concept familiarization
- General interaction approaches and best practices
- Organizational implementation context and objectives
- Expected evolution and enhancement pathways

2. **Practical interaction skills**

- Effective query and instruction formulation
- Output interpretation and validation approaches
- Exception handling and escalation procedures
- Feedback provision mechanisms
- Performance optimization techniques

3. **Role-specific application training**

- Function-specific use cases and applications
- Integration with existing workflows and processes
- Specialized tools and interaction mechanisms
- Adapting roles and responsibilities
- Performance expectations and measurement

4. **Continuous learning support**

- Self-service resources for ongoing reference
- Community forums for question resolution
- Regular update briefings on new capabilities
- Advanced technique development opportunities
- Experience sharing mechanisms across teams

Training delivery approaches

Progressive learning paths support effective skill development by introducing capabilities in stages. Learners begin with basic applications and gradually move toward more advanced ones.

This approach includes spaced practice with reinforcement and integrates relevant case applications throughout. Personalized pacing ensures the content aligns with each learner's role and experience level.

Mixed-mode delivery increases accessibility and flexibility. Formal training sessions lay the foundation, while on-demand videos and documentation provide ongoing support.

Peer mentoring and support networks encourage collaborative learning. In-workflow guidance helps learners apply knowledge in real-time, and regular practice opportunities with feedback reinforce learning.

Experiential learning places emphasis on hands-on application with realistic scenarios and explores capabilities through guided practice.

Safe environments allow for experimentation without risk, while support gradually decreases to build autonomy. Reflective exercises help consolidate insights and reinforce understanding.

The most effective training approaches recognize that collaboration with digital workers is a significant skill development journey rather than a simple tool adoption process.

Addressing Resistance and Building Buy-In

Resistance to digital workforce implementation is natural and should be expected. Change management approaches acknowledge legitimate concerns while building confidence and commitment.

Common resistance sources

1. **Job security concerns**

- Fear of replacement or redundancy
- Uncertainty about future role requirements
- Concerns about skill relevance and value
- Anxiety about adaptation ability
- Broader economic displacement worries

2. **Control and autonomy threats**

- Perceived loss of decision authority
- Concerns about judgment override
- Discomfort with process standardization
- Resistance to transparency and monitoring
- Protection of the value of domain expertise

3. **Practical implementation concerns**

- Skepticism about functionality claims
- Worry about reliability and performance
- Concerns about integration challenges
- Expectations of increased workload during transition
- Previous negative technology experience effects

4. **Identity and value challenges**

- Meaning and purpose questions for changing roles

- Status and expertise recognition concerns
- Professional identity evolution requirements
- Value contribution uncertainty
- Established relationship disruption fears

Overcoming Resistance to Change

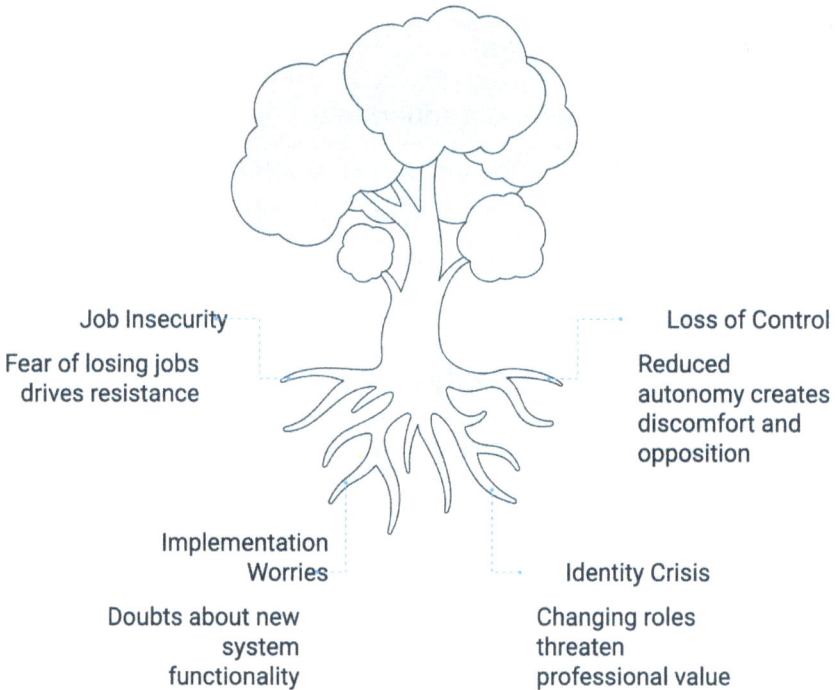

Job Insecurity
Fear of losing jobs drives resistance

Loss of Control
Reduced autonomy creates discomfort and opposition

Implementation Worries
Doubts about new system functionality

Identity Crisis
Changing roles threaten professional value

Strategies for addressing resistance

Authentic engagement involves actively listening to concerns and recognizing legitimate issues without dismissal. Leaders should be honest about both the benefits and the challenges. Balanced messaging and demonstrated interest in different perspectives help build trust and open dialogue.

Involvement and control foster a sense of ownership. This includes creating meaningful opportunities for stakeholders to provide input, making decisions within their domain, and seeing their feedback visibly reflected in outcomes. Transparency in decision-making and recognition of individual expertise further encourage participation.

Concrete demonstration of capabilities builds confidence. Hands-on experiences, early success stories, and peer testimonials help illustrate value in a tangible way. Gradual exposure to capabilities strengthens trust, making abstract benefits more real and relatable.

Clarity on individual impact clearly lays out how roles may evolve and contribute value, and identifies career development opportunities. Support for skill growth and consistent delivery on transition promises reinforces that resistance is understood and addressed.

The most effective resistance management approaches recognize that concerns typically arise from legitimate considerations rather than simple "resistance to change" and address the specific underlying issues directly.

Scaling Methodologies: Moving from Pilots to Enterprise Implementation

Once pilot implementations demonstrate value, the next challenge is scaling these initial successes across broader operations. This expansion requires systematic approaches that balance rapid value capture with sustainable implementation.

Technical Scaling Considerations

Expanding digital workforce implementations beyond initial pilots introduces complex technical challenges that must be addressed through thoughtful architecture and infrastructure approaches.

Technical scaling dimensions

1. **Infrastructure capacity expansion**

- Computing resource provisioning for increased load

- Storage capacity for expanded data requirements
- Network bandwidth for additional traffic
- Database scaling for higher transaction volumes
- Support system enhancement for broader usage

2. **Performance optimization**

- Response time maintenance under increased load
- Resource utilization efficiency improvement
- Bottleneck identification and elimination
- Parallel processing implementation, where appropriate
- Caching and optimization strategies

3. **Integration expansion**

- Connection to additional enterprise systems
- Interface standardization across implementations
- Authentication and authorization scaling
- Data flow management across the expanded scope
- API standardization and governance

4. **Security and compliance enhancement**

- Control scaling across broader implementation
- Monitoring expansion for increased activity
- Policy enforcement across diverse applications
- Audit capability expansion
- Risk management for increased exposure

Technical Scaling Dimensions

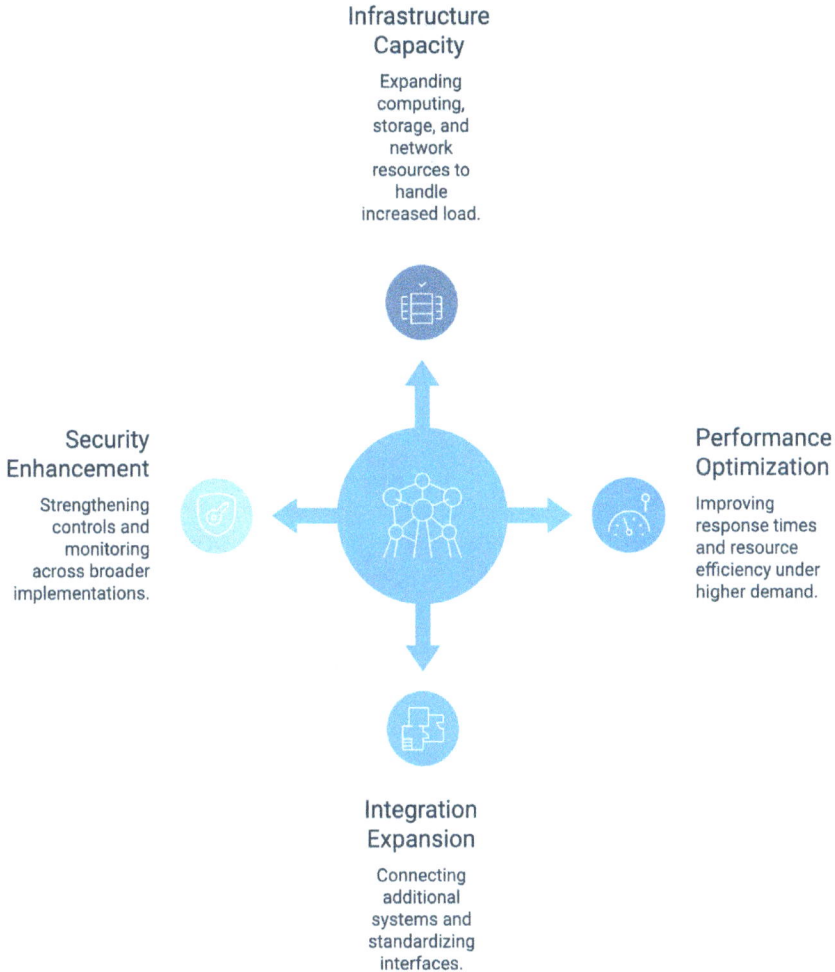

Infrastructure Capacity

Expanding computing, storage, and network resources to handle increased load.

Security Enhancement

Strengthening controls and monitoring across broader implementations.

Performance Optimization

Improving response times and resource efficiency under higher demand.

Integration Expansion

Connecting additional systems and standardizing interfaces.

Technical scaling approaches

Horizontal vs. vertical scaling decisions facilitate choosing between adding instances or enhancing existing capabilities. This includes load balancing across multiple resources, considering geographic distribution, implementing redundancy and failover mechanisms, and optimizing resource allocation.

Architecture evolution focuses on modularizing components for reuse and adopting service-oriented approaches for flexibility. It also involves choosing between centralized and distributed processing, standardizing interfaces and protocols, and packaging capabilities efficiently for deployment.

DevOps integration includes developing automated deployment pipelines, implementing continuous integration and continuous deployment (CI/CD), enhancing monitoring and alerting systems, enabling self-healing capabilities, and expanding systematic testing.

The most effective technical scaling approaches anticipate requirements beyond immediate needs, creating architectures that evolve without requiring redesign at each expansion stage.

Organizational Scaling Frameworks

Successful scaling also requires organizational frameworks that enable coordinated expansion while maintaining quality, consistency, and governance.

Organizational scaling elements

1. **Governance expansion**

- Decision rights allocation across the expanded scope
- Policy and standard development for consistency
- Oversight committee expansion and evolution
- Compliance monitoring scaling
- Value realization tracking across the portfolio

2. **Capability-building approach**

- Center of excellence establishment or expansion
- Knowledge sharing mechanisms across implementations
- Community of practice development
- Expert network creation and support

- Skill development at scale

3. **Support function evolution**

- User assistance scaling beyond pilot resources
- Training program expansion for broader audience
- Documentation development and maintenance
- Issue resolution process enhancement
- Feedback collection and processing at scale

4. **Coordination mechanism development**

- Cross-implementation synchronization
- Dependency management across initiatives
- Resource allocation across competing priorities
- Communication channels for implementation teams
- Executive alignment maintenance

Organizational Scaling Elements

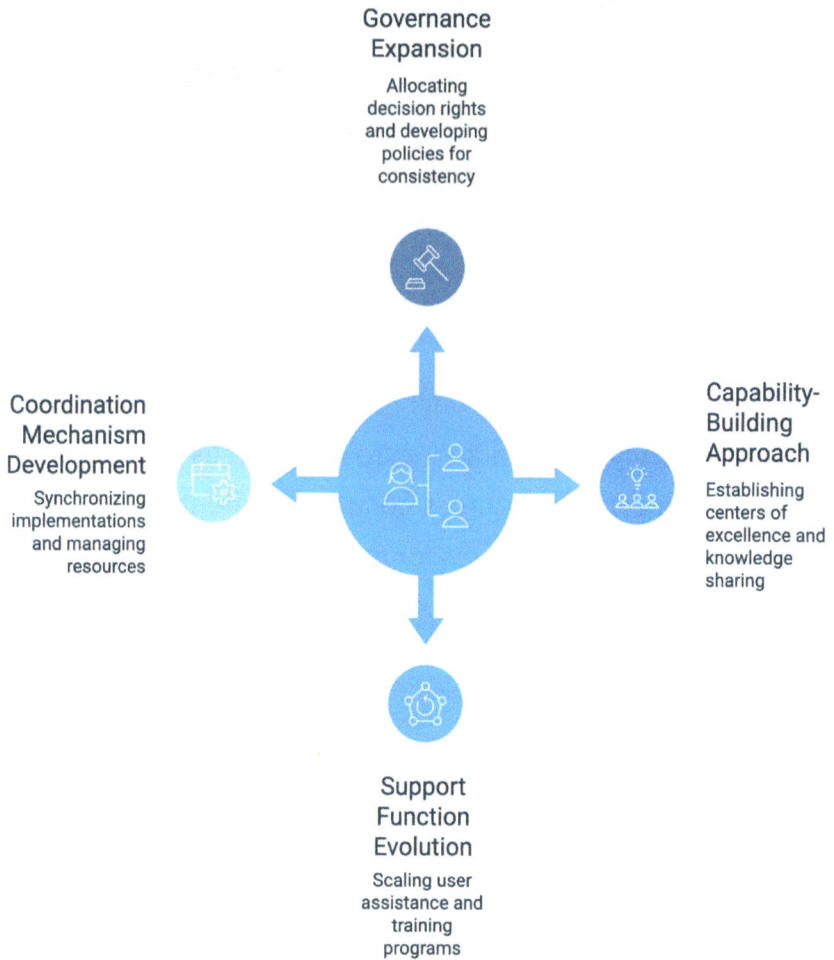

Governance Expansion

Allocating decision rights and developing policies for consistency

Coordination Mechanism Development

Synchronizing implementations and managing resources

Capability-Building Approach

Establishing centers of excellence and knowledge sharing

Support Function Evolution

Scaling user assistance and training programs

Scaling model options

1. Centralized scaling approach

- Single team responsible for all implementations
- Consistent methodology across the organization
- Centralized resource allocation and prioritization
- Standardized technology and approach
  Advantages: Consistency, efficiency, knowledge sharing

Challenges: Potential bottlenecks, distance from business

2. **Federated scaling approach**

- Central standards with distributed implementation
- Business unit implementation ownership
- Shared resources and expertise
- Coordination rather than control mechanisms
  Advantages: Business alignment, flexibility, ownership
  Challenges: Potential duplication, inconsistency

3. **Hybrid scaling models**

- Centralized platforms with distributed applications
- Core team plus embedded specialists
- Staged transition from central to distributed
- Domain-specific centers of excellence
  Advantages: Combined strengths of both approaches
  Challenges: Governance complexity, role clarity

The optimal scaling model depends on organizational structure, culture, and specific implementation characteristics. Many organizations evolve their approach as implementation maturity increases.

Resource Allocation Models for Expansion

Expanding beyond pilots requires thoughtful approaches to resource allocation that balance competing priorities while ensuring sustainable implementation.

Resource allocation considerations

1. **Investment prioritization frameworks**

- Value potential assessment methodology
- Implementation complexity and risk evaluation

- Strategic alignment determination
- Capability-building contribution
- Interdependency consideration

2. **Funding model options**

- Centralized investment pool
- Business unit funding responsibility
- Shared funding approaches
- Value-based funding with benefit sharing
- Progressive funding tied to milestone achievement

3. **Talent allocation strategies**

- Core team versus distributed expertise balance
- Build versus buy versus partner decisions for capabilities
- Knowledge transfer and rotation approaches
- Specialist versus generalist balance
- Career path development for implementation roles

4. **Vendor and partner management**

- Strategic versus tactical partner selection
- Knowledge transfer requirements and approaches
- Progressive capability internalization
- Vendor consolidation versus diversification
- Intellectual property and knowledge management

Strategic Resource Allocation Framework

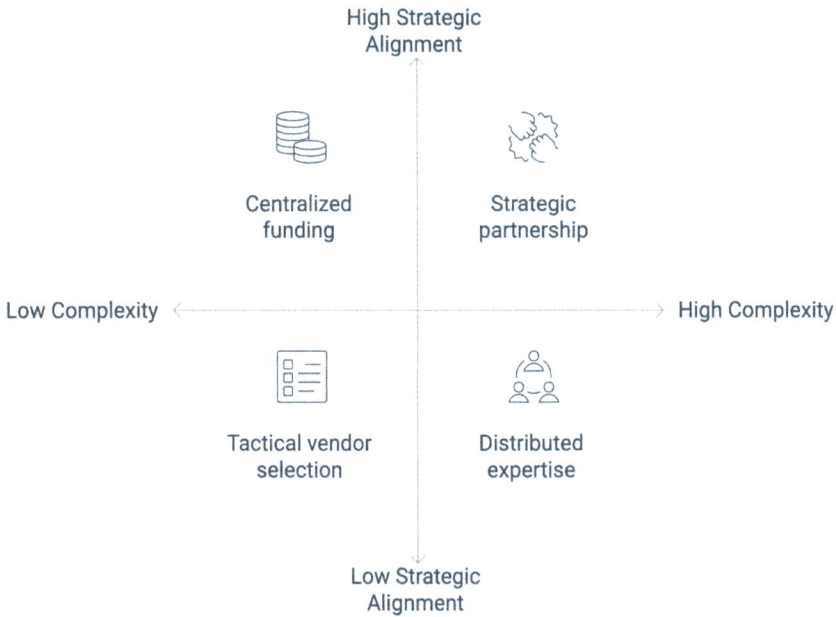

High Strategic
Alignment

Centralized
funding

Strategic
partnership

Low Complexity ←—————————————→ High Complexity

Tactical vendor
selection

Distributed
expertise

Low Strategic
Alignment

Expansion sequencing approaches

Value-driven sequencing focuses on prioritizing implementation based on potential impact. This involves early targeting of high-return opportunities, followed by progressively more complex applications.

The approach seeks to balance quick wins with longer-term strategic initiatives while managing the portfolio across different value dimensions.

Capability-building sequencing emphasizes the progressive development of implementation capabilities.

It begins with establishing foundational elements before moving to advanced applications, taking into account technical dependencies, aligning with skill development, and applying lessons learned from earlier implementations.

Change sequencing considers the organization's readiness and distributes change impact thoughtfully across groups. It aligns stakeholder support, builds momentum through visible successes, and paces the transformation to fit cultural evolution.

Effective resource allocation combines these perspectives to create balanced implementation roadmaps that deliver both immediate value and long-term capability development.

Monitoring and Maintenance: Ensuring Sustained Performance

Deployment is just the start of a digital worker's lifecycle. Sustained performance depends on comprehensive approaches to monitoring, maintenance, and continuous improvement.

Performance Monitoring Systems

Effective implementations include sophisticated monitoring capabilities that provide visibility into performance, identify emerging issues, and enable continuous optimization.

Monitoring dimension categories

1. **Technical performance monitoring**

- System availability and reliability tracking
- Response time and throughput measurement
- Resource utilization and efficiency assessment
- Error rate and exception monitoring
- Integration performance evaluation

2. **Functional performance assessment**

- Accuracy and quality measurement
- Process adherence verification
- Compliance and control effectiveness
- Output consistency evaluation

- Adaptation effectiveness to changing conditions

3. **Business impact tracking**

- Value realization measurement
- Process efficiency improvement assessment
- Quality enhancement quantification
- Capacity expansion evaluation
- User satisfaction and experience monitoring

4. **Leading indicator monitoring**

- Early warning signal identification
- Trend analysis for performance shifts
- Pattern recognition for emerging issues
- Anomaly detection across metrics
- Predictive analytics for potential problems

Monitoring Dimension Categories

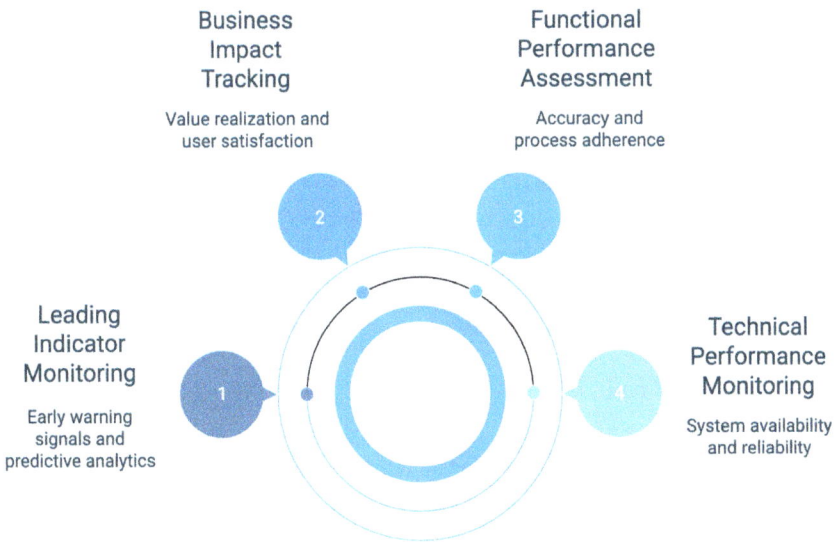

Business Impact Tracking

Value realization and user satisfaction

Functional Performance Assessment

Accuracy and process adherence

Leading Indicator Monitoring

Early warning signals and predictive analytics

Technical Performance Monitoring

System availability and reliability

Monitoring implementation approaches

Real-time dashboards provide a visual representation of key metrics, including alert mechanisms for threshold violations and drill-down capabilities for investigating issues. They are customizable for different stakeholder needs and integrate both technical and business views.

Automated analysis systems enable pattern recognition across performance data, detect and flag anomalies, and support root cause analysis. They help identify correlations across metrics and apply predictive modeling to anticipate performance trends.

Structured review processes involve a regular cadence of performance evaluations, with cross-functional participation. These reviews systematically identify and track issues, prioritize improvement opportunities, and inform decision-making forums focused on enhancement directions.

The most effective monitoring approaches combine these elements, providing real-time operational visibility and longer-term performance perspective through multiple lenses.

Issue Detection and Resolution Workflows

Even the best-designed digital workers will encounter issues that require resolution. Effective implementations include comprehensive approaches for detecting, diagnosing, and addressing these challenges.

Issue category framework

1. **Technical failures**

- System outages and availability problems
- Performance degradation and slowdowns
- Integration breakdowns with connected systems
- Resource constraints and capacity issues

- Security incidents and vulnerabilities

2. **Functional deficiencies**

- Accuracy and quality shortfalls
- Edge case handling failures
- Unexpected behavior patterns
- Context misinterpretation
- Knowledge gaps or errors

3. **User experience problems**

- Usability challenges and friction points
- Workflow integration issues
- Communication clarity failures
- Training and guidance gaps
- Adoption resistance patterns

Resolution workflow elements

1. **Detection mechanisms**

- Automated monitoring and alerting
- User reporting channels and processes
- Regular review and assessment
- Proactive testing and verification
- Early warning signal tracking

2. **Triage and prioritization**

- Impact assessment frameworks
- Urgency determination criteria
- Resolution resource allocation
- Workaround identification
- Communication requirements determination

3. Investigation and diagnosis

- Root cause analysis methodologies
- Systematic troubleshooting approaches
- Environment isolation techniques
- Reproduction procedures
- Pattern recognition across incidents

4. Resolution implementation

- Change control processes
- Testing requirements and approaches
- Deployment methodologies
- Verification procedures
- Documentation and knowledge capture

5. Prevention enhancement

- Systemic improvement identification
- Similar vulnerability assessment
- Monitoring enhancement
- Process refinement
- Knowledge sharing across teams

Resolution Workflow Elements

Detection Mechanisms	Triage and Prioritization	Investigation and Diagnosis	Resolution Implementation	Prevention Enhancement

Organizations with sophisticated issue management capabilities typically achieve higher digital worker reliability, faster resolution of problems, and more consistent performance over time.

Continuous Improvement Processes

Beyond maintaining current performance, effective digital workforce implementations include systematic approaches to continuous enhancement and evolution.

Improvement source categories

1. **Performance analysis**

- Metric trend evaluation
- Benchmark comparison
- Variation analysis
- Efficiency optimization opportunities
- Quality enhancement possibilities

2. **User feedback integration**

- Structured feedback collection
- Usage pattern analysis
- Enhancement request processing
- Pain point identification
- Satisfaction driver determination

3. **Technological advancement leverage**

- New capability evaluation
- Foundation model enhancement integration
- Emerging technique assessment
- Architecture evolution opportunities
- Tool and platform advancement incorporation

4. **Organizational learning application**

- Cross-implementation insight sharing
- Best practice identification and transfer

- Failure pattern recognition
- Success factor determination
- Knowledge community development

Exploring Improvement Source Categories

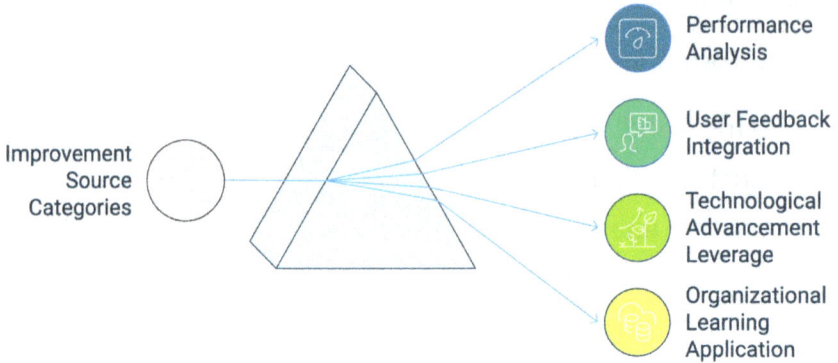

Improvement Source Categories

Performance Analysis

User Feedback Integration

Technological Advancement Leverage

Organizational Learning Application

Improvement implementation approaches

Agile enhancement cycles focus on regular release planning and prioritization, using time-boxed sprints to deliver incremental capabilities. Continuous integration methods support rapid deployment, while ongoing feedback guides refinement.

Structured innovation programs support dedicated exploration initiatives and experimentation frameworks. These include pilot testing of new approaches and the controlled introduction of emerging capabilities within sandbox environments for evaluation.

User co-creation emphasizes collaborative design sessions and usability testing. Early adopter programs allow for real-world input, with user feedback shaping feature prioritization and closing the loop for continuous improvement.

The most effective improvement processes balance immediate enhancement of existing capabilities with longer-term evolution toward new possibilities, creating sustainable advancement paths for digital workforce value.

Conclusion: From Implementation to Transformation

The successful deployment and scaling of digital workers require integrated approaches that address both technical and organizational dimensions. Leading organizations recognize that implementation is not merely a project but a transformation journey that evolves over time.

Several key principles distinguish successful approaches:

1. **Balance speed and sustainability**: Finding the right pace that captures value quickly while building foundations for long-term success rather than sacrificing either for the other.

2. **Integrate technical and organizational change**: Recognizing that successful implementation requires simultaneous evolution of technology, processes, skills, and mindsets rather than focusing exclusively on any single dimension.

3. **Build learning organizations**: Creating mechanisms for systematic knowledge capture, sharing, and application rather than allowing each implementation to operate in isolation.

4. **Maintain strategic alignment**: Continuously connecting implementation decisions to broader organizational objectives rather than allowing technology-driven mission creep.

5. **Emphasize human experience**: Designing implementation approaches around effective human-digital collaboration rather than focusing exclusively on technology capabilities.

As digital workforces continue to advance, the organizations that thrive will be those that develop sophisticated deployment and scaling capabilities, transforming initial successes into enterprise-wide transformation through methodical expansion and continuous evolution.

PART
THREE
MANAGEMENT AND GOVERNANCE

CHAPTER 7:
MANAGING DIGITAL WORKERS

TL;DR:

- Digital workers require new supervision models, from direct human-in-the-loop oversight for critical functions to monitoring-based human-on-the-loop approaches for routine operations.

- Effective management includes thoughtful incentive design that shapes digital worker behavior toward organizational objectives through appropriate reinforcement approaches.

- Comprehensive performance management combines multidimensional measurement, systematic quality assurance, and continuous improvement methodologies.

- Building effective human-digital teams requires appropriate structures, clear communication protocols, and systematic trust-building approaches.

- Organizations that develop sophisticated digital workforce management capabilities create sustainable advantages through superior utilization of these powerful new resources.

Effective Digital Workforce Management

Once digital workers are deployed, a new management challenge emerges. Organizations must develop approaches for effectively supervising, directing, and optimizing agentic systems as ongoing operational resources rather than one-time implementation projects.

Supervision Models: Approaches to Digital Workforce Oversight

The supervision of digital workers is a different challenge from traditional human management. Organizations must develop new models that provide appropriate oversight while leveraging the unique capabilities of agentic systems.

Human-in-the-Loop vs. Human-on-the-Loop Models

Different digital worker applications require different levels of human involvement, ranging from direct supervision of each action to periodic review of autonomous operations.

Human-in-the-loop supervision

This model positions humans as active participants in the digital worker's operational flow. Humans review, approve, or modify system outputs before they take effect.

Direct approval mechanisms ensure human oversight by requiring review of recommendations before implementation and explicit approvals for consequential actions. Decision checkpoints are placed at critical stages, allowing modification of outputs before finalization. Transparency is maintained through clear visibility into the system's reasoning and sources.

Key application contexts: High-stakes decisions with significant consequences

- Novel or rapidly changing environments

- Complex judgment situations with ethical dimensions
- Regulatory environments requiring human accountability
- Early deployment phases, while building trust and reliability

Implementation considerations: Balance between oversight value and efficiency impact

- Clear responsibility delineation for decision quality
- Workload management for human reviewers
- Decision criteria transparency for both humans and systems
- Progressive adaptation as experience and trust increase

Human-on-the-loop supervision

This model positions humans as monitors and exception handlers for largely autonomous digital worker operations, intervening only when necessary.

Monitoring and exception handling involve real-time oversight of system performance with alert mechanisms to flag unusual patterns or outcomes. Systems are designed with intervention capabilities when necessary, supported by periodic sample reviews of routine operations. Threshold-based escalation mechanisms help ensure timely responses to specified conditions.

Key application contexts: High-volume, routine operations with established patterns

- Well-understood domains with clear success criteria
- Lower-stakes decisions with limited potential harm
- Mature implementations with demonstrated reliability
- Areas with clear performance metrics and monitoring

Implementation considerations: Alert design to prevent oversight fatigue

Appropriate sampling approaches for quality control

Escalation threshold calibration

Record-keeping for accountability

Continuous improvement mechanisms

The choice between these models, or the implementation of hybrid approaches that vary by process stage or decision type, should be based on a thoughtful analysis of risk and efficiency. Factors such as capability maturity and regulatory requirements should also inform the decision, rather than defaulting to either extreme.

Escalation Protocols and Intervention Triggers

Effective digital worker management requires clear protocols for when and how humans should intervene in otherwise autonomous operations.

Escalation trigger categories

1. **Confidence-based triggers**: System self-assessment of capability limitations

- Uncertainty thresholds for different decision types
- Probability distribution analysis for potential outcomes
- Consistency evaluation across multiple approaches
- Novel pattern recognition requiring judgment

2. **Impact-based triggers**: Consequence magnitude assessment

- Irreversibility of potential actions
- Stakeholder impact evaluation
- Reputation and trust implications
- Financial or operational risk thresholds

3. **Pattern-based triggers**: Deviation from historical performance patterns

- Unusual input or environmental conditions
- Request characteristics outside normal parameters
- Sequence or timing anomalies
- Multiple related exceptions within the timeframe

4. **Compliance-based triggers**: Regulatory requirement adherence

- Policy compliance verification
- Ethical boundary condition detection
- Documentation completeness assessment
- Approval requirement identification

Escalation Trigger Categories

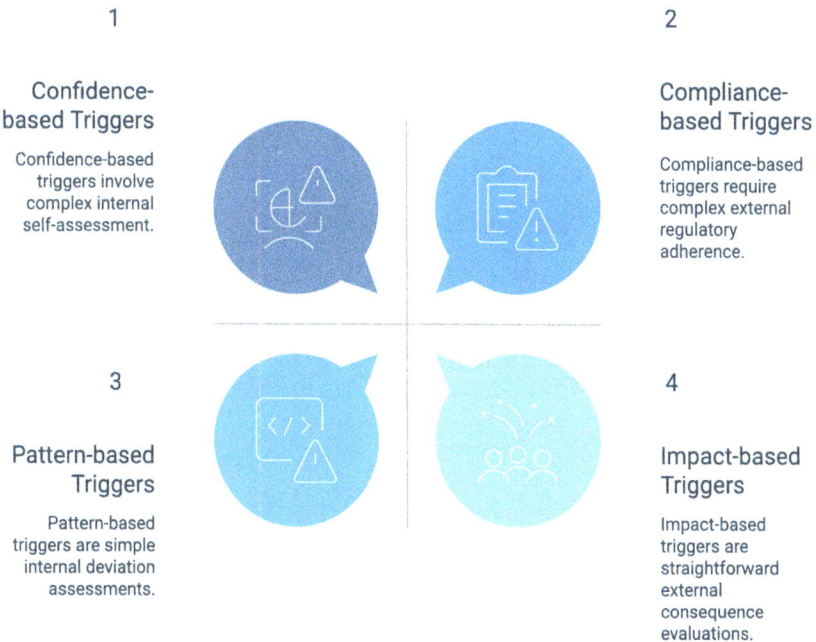

1

Confidence-based Triggers

Confidence-based triggers involve complex internal self-assessment.

2

Compliance-based Triggers

Compliance-based triggers require complex external regulatory adherence.

3

Pattern-based Triggers

Pattern-based triggers are simple internal deviation assessments.

4

Impact-based Triggers

Impact-based triggers are straightforward external consequence evaluations.

Escalation protocol design elements

1. **Routing and responsibility frameworks**: Clear designation of escalation recipients

- Tiered response based on issue characteristics
- Backup mechanisms for unavailable responders
- Cross-functional routing for complex issues
- Service level agreements for response times

2. **Context provision mechanisms**: Relevant information package assembly

- Decision history and rationale capture
- Reference to similar previous situations
- Action option development
- Consequence analysis for alternatives

3. **Resolution documentation**: Decision recording for accountability

- Rationale capture for future reference
- Learning integration from resolutions
- Pattern analysis across escalations
- Improvement opportunity identification

Sophisticated escalation frameworks help achieve an optimal balance between autonomous operation and appropriate human intervention. Organizations then gradually shift this balance as system capabilities mature and human confidence increases.

Performance Review Methodologies

Just as human employees benefit from structured performance evaluation, digital workers require systematic review processes to ensure they continue to meet organizational needs and expectations.

Review dimension categories

1. **Operational performance assessment**: Accuracy and quality measurements

- Efficiency and throughput evaluation
- Reliability and consistency assessment
- Response time and availability analysis
- Exception frequency and handling effectiveness

2. **Business impact evaluation**: Value delivery against objectives

- Return on investment realization
- Process improvement contribution
- User satisfaction and adoption
- Competitive advantage creation

3. **Risk and compliance verification**: Policy adherence confirmation

- Security control effectiveness
- Privacy protection assessment
- Ethical alignment evaluation
- Regulatory requirement compliance

4. **Evolution and improvement analysis**: Learning and adaptation effectiveness

- Performance trend evaluation
- Enhancement implementation assessment
- Emerging capability needs identification
- Strategic alignment verification

Comprehensive Review Framework

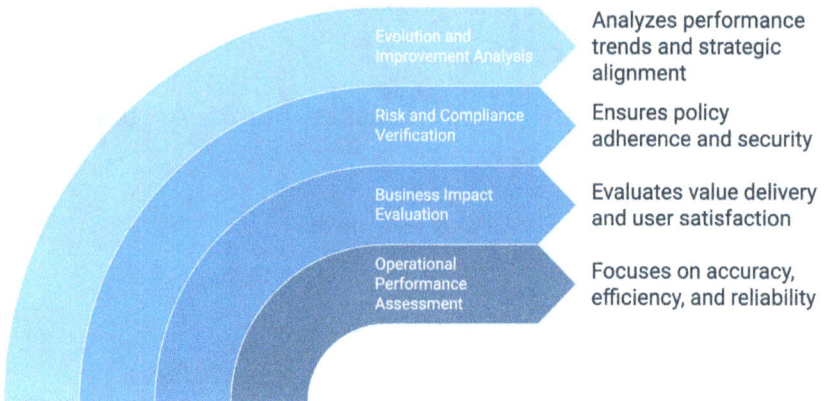

Evolution and Improvement Analysis	Analyzes performance trends and strategic alignment
Risk and Compliance Verification	Ensures policy adherence and security
Business Impact Evaluation	Evaluates value delivery and user satisfaction
Operational Performance Assessment	Focuses on accuracy, efficiency, and reliability

Review process implementation

1. **Cadence and structure**: Regularly scheduled comprehensive reviews

- Continuous monitoring with threshold alerts
- Milestone-based evaluation for major changes
- Multi-level review hierarchy (operational to strategic)
- Cross-functional assessment participation

2. **Methodology components**: Quantitative metric evaluation

- Qualitative feedback integration
- Sample case examination
- Comparative benchmark analysis
- Future requirement assessment

3. **Outcome integration**: Improvement prioritization framework

- Enhancement resource allocation
- Capability evolution planning
- Operational adjustment implementation

- Stakeholder communication approach

Effective performance review processes not only verify current functioning but also drive continuous improvement and strategic alignment of digital worker capabilities over time.

Incentive Alignment: Creating Effective Reward Systems

Although digital workers lack the intrinsic and extrinsic motivations of humans, their behavior can and should be shaped through thoughtfully designed incentive structures that align their operations with organizational objectives.

Reinforcement Learning Approaches for Behavior Shaping

Many digital workers incorporate capabilities to learn from feedback and adjust their behavior accordingly. Organizations can leverage these capabilities through systematic reinforcement approaches.

Reinforcement mechanism types

1. **Explicit feedback systems**: Direct quality ratings of outputs

- Binary acceptance or rejection signals
- Graduated scoring of response value
- Comparative ranking among alternatives
- Detailed feedback with improvement guidance

2. **Implicit signal utilization**: User behavior pattern analysis

- Adoption and utilization tracking
- Selection among the provided options
- Time spent with different outputs
- Follow-up action monitoring

3. **Outcome-based reinforcement**: End result measurement

- Process completion success
- Downstream impact assessment
- Long-term value realization
- Strategic objective contribution

Which reinforcement mechanism type should be implemented?

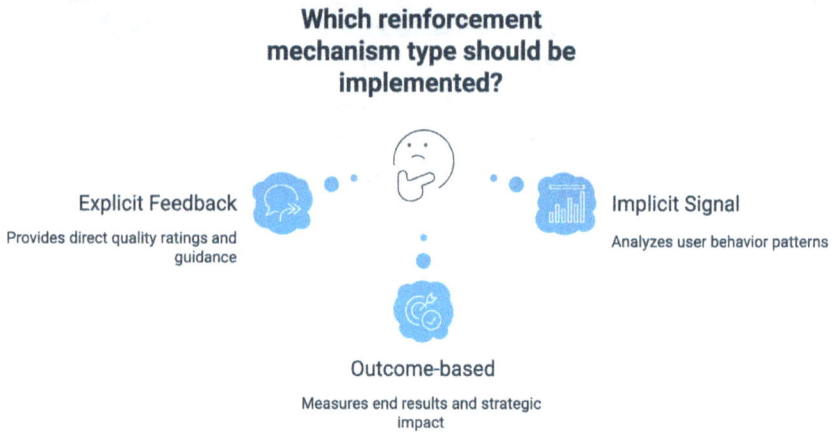

Explicit Feedback
Provides direct quality ratings and guidance

Implicit Signal
Analyzes user behavior patterns

Outcome-based
Measures end results and strategic impact

Implementation considerations

Signal quality management: Effective implementation requires managing signal quality through bias identification and mitigation, noise reduction in feedback, and ensuring consistency across evaluators. It also involves applying context-appropriate weighting and balancing immediate versus delayed signals.

Learning rate calibration: Controlling adaptability speed is crucial, with attention to balancing stability and responsiveness. Strategies include using historical weighting approaches, handling novel situations appropriately, and putting safeguards in place to prevent incorrect learning.

Oversight and verification: Monitoring learning patterns involves regular audits of behavior, detecting performance drift, and identifying triggers for intervention. When necessary, reset mechanisms should be in place to restore appropriate functioning.

Well-designed reinforcement approaches enable digital workers to continuously improve their alignment with organizational needs while maintaining appropriate boundaries and safeguards.

Evaluation Criteria Design

The behavior of digital workers is significantly shaped by the criteria used to evaluate their performance. Thoughtful design of these criteria ensures alignment with true organizational objectives rather than proxy metrics that may create unintended consequences.

Criteria development principles

Outcome focus vs. process adherence: Criteria should emphasize results rather than rigid adherence to methods where appropriate. Flexibility must be allowed for novel approaches that achieve objectives.

Goal-oriented metrics can encourage innovation. The right balance between focusing on "how" versus "what" should be context-dependent, with a clear distinction between required controls and merely preferred methods.

Multidimensional assessment: Effective criteria should cover multiple value dimensions and offer explicit trade-off guidance when dimensions conflict.

Weighting should reflect strategic importance, with a balanced view of both short- and long-term impacts. Both quantitative and qualitative factors need to be included for a comprehensive evaluation.

Stakeholder perspective integration: All affected parties should be considered, with an emphasis on end-user experience.

Criteria should align with broader business objectives while also incorporating risk and compliance perspectives. Balancing strategic priorities with operational realities ensures more practical and inclusive outcomes.

Common evaluation categories

1. **Effectiveness dimensions**: Task completion success rate

- Accuracy and precision
- Completeness and comprehensiveness
- Relevance and appropriateness
- Value addition beyond basic requirements

2. **Efficiency dimensions**: Resource utilization optimization

- Time and effort minimization
- Process streamlining contribution
- Automation level appropriateness
- Cost effectiveness relative to alternatives

3. **Experience dimensions**: User satisfaction and perception

- Interaction quality and naturalness
- Friction reduction contribution
- Accessibility and inclusivity
- Trust-building effectiveness

4. **Evolution dimensions**: Learning and improvement over time

- Adaptation to changing conditions
- Evidence of knowledge expansion
- Improvement of novel situation handling
- Self-optimization capability

Exploring Evaluation Categories

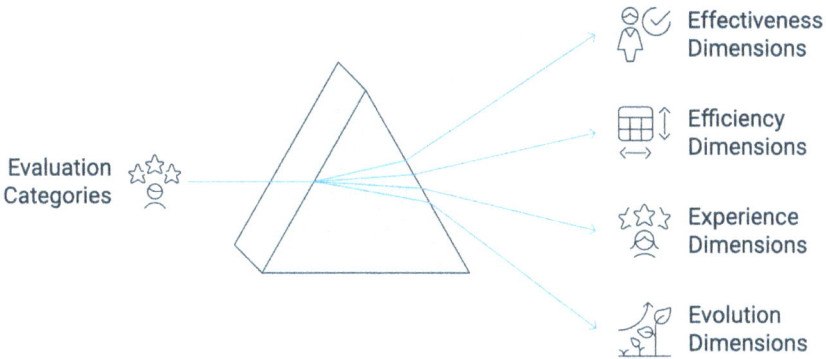

Evaluation Categories → Effectiveness Dimensions, Efficiency Dimensions, Experience Dimensions, Evolution Dimensions

Organizations with sophisticated evaluation frameworks typically achieve better alignment between digital worker behavior and true organizational objectives, avoiding the common pitfall of optimizing for easily measured metrics at the expense of more important but less quantifiable outcomes.

Balancing Multiple Objectives

Digital workers often operate in complex environments with multiple, sometimes competing objectives. Effective management requires clear frameworks for navigating these trade-offs.

Common objective tension areas

1. **Speed vs. quality**: Response time expectations

- Thoroughness requirements
- Verification level appropriate to context
- Progressive delivery approaches
- Situation-specific balance calibration

2. **Standardization vs. personalization**: Process consistency requirements

- Accommodation for individual needs

- Procedural compliance obligations
- Experience customization expectations
- Context-dependent flexibility parameters

3. **Autonomy vs. human judgment**: Decision authority allocation

- Escalation threshold determination
- Autonomy boundary definition
- Override mechanism design
- Progressive authority expansion approach

4. **Exploration vs. exploitation**: New approach experimentation allowance

- Proven method utilization expectations
- Innovation encouragement parameters
- Reliability requirements
- Learning opportunity provision

Trade-off management approaches

Explicit prioritization frameworks: Establishing clear hierarchies for conflicting objectives is key. This includes defining context-dependent priority shifting, being transparent with decision criteria, specifying boundary conditions, and presenting override justifications.

Situational calibration mechanisms: Balancing trade-offs often requires context-specific calibration. Factors like the level of risk, stakeholder impact, and uncertainty inform how priorities are weighed. Additionally, adjusting for different time horizons allows the approach to remain adaptive and relevant.

Hybrid approaches: Segmenting responsibilities between human and digital workers, staging processes with varying emphasis, and exploring parallel paths with final human selection can create a balanced system. Continuous recalibration based on outcomes ensures the approach remains responsive and aligned.

Robust approaches to objective balancing typically help organizations achieve more nuanced digital worker behavior that appropriately navigates complex trade-offs rather than rigidly prioritizing single dimensions regardless of context.

Performance Management: Measuring and Improving Digital Worker Effectiveness

Beyond basic supervision, organizations must implement comprehensive performance management approaches that drive continuous improvement in digital worker capabilities and impact.

Key Performance Indicators (KPIs) for Digital Workers

Effective performance management begins with appropriate metrics that capture the multiple dimensions of digital worker contribution to organizational objectives.

KPI category framework

1. **Operational metrics**: Task completion volume and throughput

- Error rates and quality levels
- Processing time and efficiency
- Availability and reliability
- Exception handling effectiveness

2. **Business impact metrics**: Cost reduction realization

- Revenue contribution
- Customer experience improvement
- Employee productivity impact

- Process quality enhancement

3. **Strategic contribution metrics**: Enabling innovation

- Competitive differentiation
- Business model evolution support
- Strategic initiative advancement
- Organizational capability enhancement

4. **Learning and improvement metrics**: Performance trend over time

- Adaptation to changing conditions
- Knowledge expansion evidence
- Self-optimization capability
- Enhancement implementation effectiveness

KPI Category Comparison

	Operational	Business Impact	Strategic Contribution	Learning & Improvement
Focus	Task completion volume	Cost reduction realization	Enabling innovation	Performance trend over time
Quality	Error rates and quality	Customer experience improvement	Competitive differentiation	Adaptation to changing conditions
Efficiency	Processing time and efficiency	Revenue contribution	Business model evolution support	Knowledge expansion evidence
Reliability	Availability and reliability	Employee productivity impact	Strategic initiative advancement	Self-optimization capability
Handling	Exception handling effectiveness	Process quality enhancement	Organizational capability enhancement	Enhancement implementation effectiveness

Implementation considerations

Measurement system design involves integrating data collection mechanisms into workflows and determining the appropriate frequency of measurement. It also includes establishing baselines for comparison, identifying benchmarks where available, and visualizing data in formats suitable for different stakeholders.

Context consideration ensures the metrics account for environmental factors, such as seasonal or cyclical patterns, and normalize input variations to allow fair comparisons. It also requires identifying external influences that could skew performance data.

As organizations learn and grow, metrics must adapt. This means refining existing KPIs, progressively raising performance expectations, shifting emphasis as digital capabilities mature, adding new dimensions when relevant, and retiring metrics that have become obsolete.

The most effective KPI frameworks combine these perspectives to create a balanced view of digital worker performance across multiple timeframes and stakeholder priorities.

Quality Assurance Frameworks

Beyond measurement, organizations must implement systematic approaches to ensure digital workers consistently deliver appropriate quality outcomes.

Quality dimension categories

1. **Functional quality**: Accuracy and correctness

- Completeness and comprehensiveness
- Consistency across similar situations
- Appropriate handling of edge cases
- Alignment with requirements and specifications

2. **Experience quality**: Usability and interaction effectiveness

- Response appropriateness and helpfulness
- Communication clarity and accessibility

- Expectation alignment
- Trust-building effectiveness

3. **Security and compliance quality**: Data protection and privacy compliance

- Policy adherence and control effectiveness
- Auditability and traceability
- Authorization boundary respect
- Ethical alignment and social responsibility

Foundations of Quality

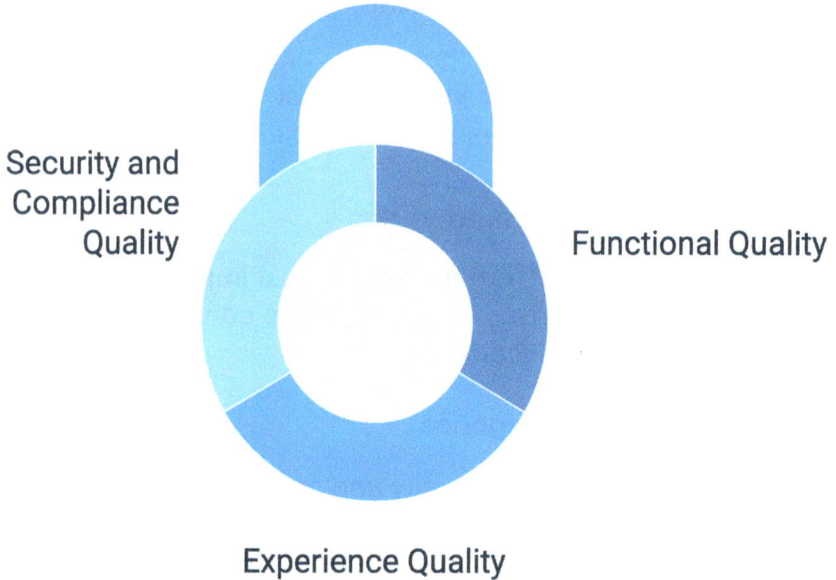

Security and Compliance Quality

Functional Quality

Experience Quality

Quality assurance methodologies

Proactive quality approaches include systematic testing across a wide range of scenarios, thorough pre-deployment verification, and adversarial evaluation to identify weaknesses.

Simulations in controlled environments help assess behavior under varied conditions, while progressive exposure strategies are used based on growing confidence in quality.

Operational quality control involves real-time monitoring for quality indicators, implementing sampling approaches for human review, and using automated verification wherever feasible.

User feedback plays a vital role in identifying issues early, and exception pattern analysis helps pinpoint recurring problems.

Continuous quality improvement includes performing root cause analysis of quality issues, prioritizing systematic enhancements, and verifying closure of feedback loops.

Monitoring performance trends enables early detection of quality drift, while benchmarking helps establish and maintain high standards.

Organizations with robust quality assurance frameworks typically achieve higher reliability, greater user confidence, and more consistent value delivery from their digital workforce investments.

Continuous Improvement Methodologies

Digital workers possess the unique capability to continuously improve through learning and adaptation. Effective management approaches leverage this capability through systematic improvement methodologies.

Improvement source categories

1. **Direct performance feedback**: User evaluation and ratings

- Output quality assessment
- Process efficiency analysis
- Exception frequency and patterns
- Business impact measurement

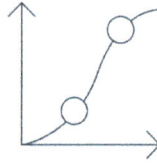

2. **Comparative analysis**: Peer performance benchmarking

- Best practice identification
- Variation analysis across instances
- Performance distribution examination
- Gap assessment against targets

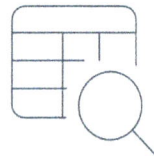

3. **Advanced analytics**: Pattern recognition across operations

- Predictive modeling for enhancement impact
- Simulation for improvement testing
- Multivariate optimization techniques
- A/B testing for alternative approaches

Improvement Source Categories

Direct Performance Feedback	Comparative Analysis	Advanced Analytics
User evaluation and ratings for output quality and process efficiency.	Peer performance benchmarking to identify best practices and assess gaps.	Pattern recognition across operations using predictive modeling and simulation.
1	2	3

Improvement implementation approaches

Systematic enhancement processes involve setting a regular cadence for improvement cycles, supported by a prioritization framework to identify and select the most impactful opportunities. This includes careful planning and execution of implementation steps, knowledge sharing, and methodologies to verify the impact.

Capability evolution management includes integrating updates to foundation models, deploying new features through defined methodologies, ensuring backward compatibility, expanding capabilities progressively, and managing the transition from legacy functionalities.

Collaborative improvement: involves leveraging human-AI partnerships to drive enhancements, integrating expert input through integration mechanisms, forming cross-functional improvement teams, creating channels for user community contributions, and fostering collaboration with partners and vendors.

Organizations that implement sophisticated improvement methodologies typically achieve accelerating value from their digital workforce investments over time rather than the degradation commonly seen with traditional technology implementations.

Team Integration: Building Effective Human-AI Teams

The greatest value from digital workers often comes not from isolated automation but from effective collaboration with human team members. Building these integrated teams requires thoughtful approaches to structure, communication, and trust development.

Team Structure Best Practices

How human-digital teams are organized shapes their effectiveness, innovation potential, and ability to leverage the unique strengths of each participant type.

Structural model options

1. **Digital assistant model**: Digital workers support individual human team members

- One-to-one or one-to-few relationships
- Human direction and decision authority
- Digital focus on information, analysis, and execution support
- Human focus on judgment, relationships, and strategy

2. **Peer collaboration model**: Digital workers operate as team members alongside humans

- Task allocation based on comparative advantage
- Shared responsibility for outcomes
- Complementary capability leveraging
- Mutual support and enhancement

3. **Augmented team model**: Digital capabilities embedded throughout team operations

- Fluid boundary between human and digital contributions
- Contextual shifting of responsibility
- Continuous capability enhancement focus
- Progressive evolution of relationship

Implementation considerations

1. **Role and responsibility definition**: Clear delineation of accountabilities

- Decision rights allocation
- Escalation and exception path specification
- Overlap management for shared domains
- Evolution pathways as capabilities mature

2. **Team size and composition**: Optimal human-digital ratio determination

- Skill complement identification
- Diversity considerations for both types
- Specialization versus generalization balance
- Backup and redundancy approaches

3. **Leadership and coordination**: Human leadership role definition

- Coordination mechanism design
- Performance management responsibility
- Development and growth facilitation
- Culture and relationship building

Effective Human-AI Team Dynamics

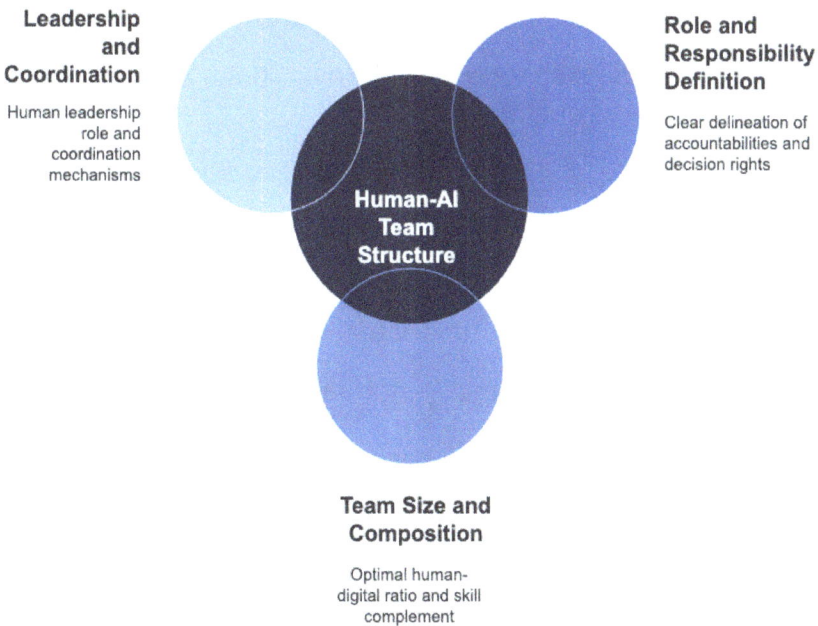

Leadership and Coordination

Human leadership role and coordination mechanisms

Role and Responsibility Definition

Clear delineation of accountabilities and decision rights

Human-AI Team Structure

Team Size and Composition

Optimal human-digital ratio and skill complement

The most effective structural approaches typically balance clear accountability with collaborative flexibility. They create teams that leverage the strengths of both human and digital members while compensating for their respective limitations.

Communication Protocols and Coordination Mechanisms

Effective human-digital teams require clear communication protocols and coordination mechanisms that enable seamless collaboration across the human-machine boundary.

Communication protocol elements

1. **Interaction modality design**: Channel selection appropriate to context

- Multimodal communication options
- Accessibility and inclusivity considerations
- Synchronous versus asynchronous approaches
- Private versus shared communication spaces

2. **Information exchange standards**: Common terminology and definition establishment

- Context sharing expectations
- Update frequency and triggers
- Documentation and record-keeping requirements
- Transparency level appropriate to situation

3. **Feedback mechanisms**: Performance input approaches

- Improvement suggestion channels
- Issue reporting procedures
- Recognition and acknowledgment practices
- Learning integration from communication

Communication Strategies Comparison

Characteristic	Channel Selection	Information Exchange	Feedback Mechanisms
Context	Appropriate choice	Terminology use	Performance input
Modality	Multimodal options	Context sharing	Improvement suggestions
Timing	Synchronous or asynchronous	Update frequency	Issue reporting
Space	Private or shared	Documentation needs	Recognition practices
Other	Accessibility focus	Transparency level	Learning integration

Coordination approach categories

1. **Process-based coordination**: Structured workflows with clear handoffs

- Stage gates and checkpoints
- Progress tracking mechanisms
- Dependency management approaches
- Exception handling procedures

2. **Shared awareness coordination**: Common information space maintenance

- Status visibility mechanisms
- Intent signaling approaches
- Context synchronization processes
- Environmental change notification

3. **Adaptive coordination**: Dynamic role allocation based on situation

- Real-time adjustment to changing conditions
- Opportunity identification and leverage
- Emergent pattern recognition and response
- Continuous learning from coordination experiences

Advanced communication and coordination approaches translate to higher performance from integrated human-digital teams, particularly in complex and dynamic environments.

Trust-Building Methodologies

Perhaps the most critical foundation for effective human-digital teams is trust—the confidence that both human and digital participants will perform appropriately, communicate honestly, and support team objectives.

Trust dimension categories

1. **Competence trust**: Belief in the capability to perform effectively

- Confidence in appropriate skill application
- Expectation of quality outcomes
- Recognition of improvement over time
- Understanding of limitation boundaries

2. **Reliability trust**: Expectation of consistent performance

- Confidence in availability when needed
- Belief in commitment to promises
- Predictability of behavior patterns
- Dependability under pressure or constraints

3. **Transparency trust**: Confidence in honest communication

- Belief in appropriate disclosure
- Understanding of reasoning and rationale
- Visibility into underlying processes
- Clarity about confidence levels and uncertainty

4. **Value alignment trust**: Confidence in shared objectives

- Belief in common priorities
- Understanding of ethical frameworks
- Expectation of an appropriate trade-off navigation
- Recognition of consistent value demonstration

Trust dimensions range from individual to shared values.

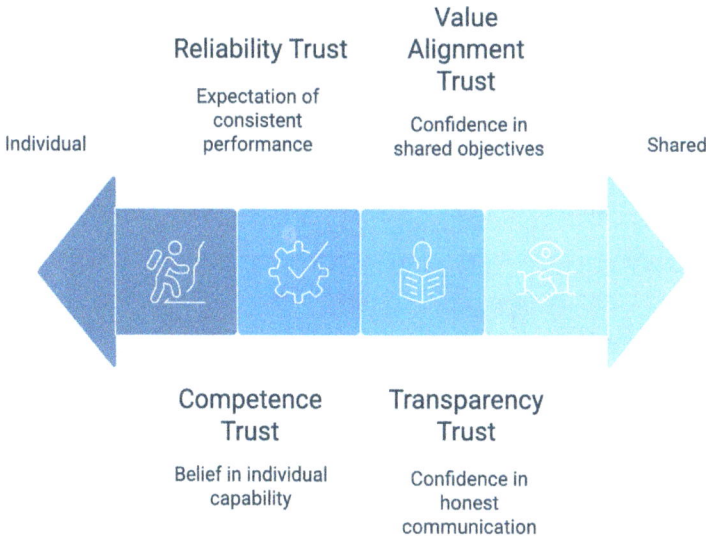

Individual → Shared

Reliability Trust — Expectation of consistent performance

Value Alignment Trust — Confidence in shared objectives

Competence Trust — Belief in individual capability

Transparency Trust — Confidence in honest communication

Trust development approaches

Progressive exposure focuses on gradually increasing the use and responsibility of digital workers. It begins with a staged introduction in environments with lower risk, allowing controlled testing.

Capabilities are demonstrated before the system is relied on for critical tasks. Over time, shared history builds experience, and confidence grows as performance is consistently verified.

Expectation management ensures that stakeholders have a clear and realistic understanding of what the system can do. This involves being transparent about limitations, signaling confidence levels clearly, acknowledging uncertainty when necessary, and providing honest assessments of performance.

Explanation and understanding are essential for trust and adoption. This includes offering context-appropriate transparency into reasoning, making the rationale behind decisions visible, disclosing information sources, and explaining processes at a suitable level of detail.

Clarifying the system's learning approach also helps users understand its behavior and limitations.

Organizations that invest in systematic trust-building typically achieve higher adoption, more effective collaboration, and greater value realization from their human-digital teams than those focusing exclusively on technical performance.

Conclusion: The Evolving Discipline of Digital Workforce Management

Managing digital workers is a new discipline that combines elements of traditional management, technology governance, and system operation while introducing novel challenges and opportunities.

Several key principles distinguish successful approaches to digital workforce management:

1. **Balance autonomy and oversight**: Finding the right level of human involvement for different contexts, gradually

increasing autonomy as capabilities and trust develop rather than defaulting to either extreme.

2. **Align incentives and objectives**: Creating clear frameworks for what digital workers should optimize for, including explicit guidance for navigating trade-offs rather than single-dimension targets that create unintended consequences.

3. **Measure multidimensionally**: Implementing comprehensive performance measurement approaches that capture the full range of value creation rather than focusing on easily quantified but potentially misleading metrics.

4. **Enable continuous improvement**: Leveraging the unique capability of digital workers to learn and adapt through systematic feedback integration and enhancement processes.

5. **Build integrated teams**: Designing structures, protocols, and practices that enable effective human-digital collaboration rather than treating digital workers as isolated automation tools.

The sophistication of digital workforce management approaches will increasingly differentiate organizations that capture their full potential from those that achieve only marginal benefits. Investing in this new management discipline, while less visible than the technology itself, often determines the ultimate success of digital workforce initiatives.

CHAPTER 8:
RISK MANAGEMENT AND COMPLIANCE

TL;DR:

- Digital workers introduce novel risk profiles requiring systematic assessment across security, operational, and reputational dimensions through integrated frameworks that capture interaction effects.

- The regulatory landscape for digital workers spans multiple domains with significant variation by geography and industry, requiring comprehensive compliance strategies that adapt to evolving requirements.

- Ethical implementation requires structured approaches to bias mitigation, fairness assessment, transparency, and responsible innovation governance to ensure digital workers operate in alignment with organizational and societal values.

- Even with robust prevention, failures will occur, necessitating thoughtful fallback design, business continuity planning, and recovery protocols to maintain operational resilience when disruptions inevitably happen.

- Successful organizations balance innovation and protection by integrating risk management throughout the digital worker lifecycle, calibrating controls proportionally, and continuously evolving their approaches as both capabilities and threats advance.

Integrated Risk and Compliance Management

As organizations increasingly deploy digital workers for business-critical functions, systematic approaches to risk management and compliance become essential.

The unique characteristics of agentic AI systems introduce novel risks alongside traditional concerns, requiring thoughtful frameworks for identification, assessment, mitigation, and governance.

Risk Assessment: Identifying and Evaluating Digital Workforce Risks

Effective risk management begins with comprehensive assessment processes that systematically identify, categorize, and evaluate potential risks across multiple dimensions.

Security Risk Evaluation Frameworks

Digital workers often have extensive access to sensitive systems and information, creating unique security risk profiles that must be systematically evaluated and managed.

Key security risk categories

1. **Authentication and authorization risks**

- Digital identity compromise
- Permission escalation vulnerabilities
- Credential management weaknesses
- Session hijacking opportunities
- Boundary control failures

2. **Data protection vulnerabilities**

- Information exfiltration pathways
- Unintended data exposure

- Processing security weaknesses
- Transit protection gaps
- Storage security limitations

3. **System integrity threats**

- Unauthorized modification possibilities
- Input validation weaknesses
- Configuration vulnerability exploitation
- Supply chain compromise risks
- Dependency security concerns

4. **Novel attack vectors**

- Prompt injection vulnerabilities
- Training data contamination
- Adversarial manipulation techniques
- Model extraction attacks
- Inference optimization exploits

Security Risk Categories

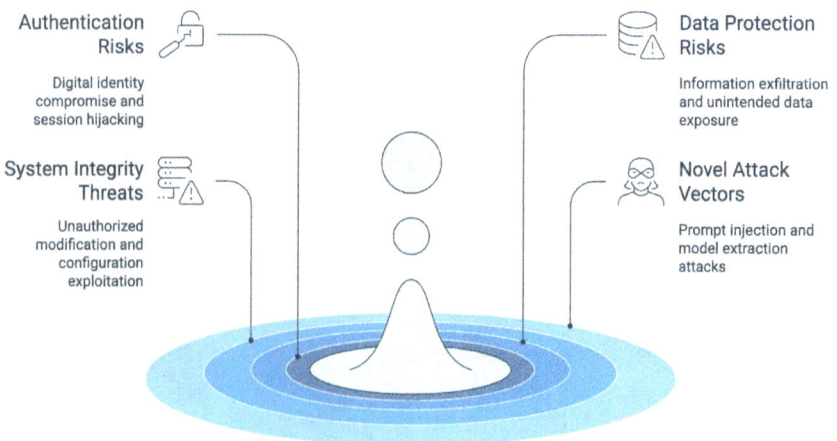

Authentication Risks
Digital identity compromise and session hijacking

Data Protection Risks
Information exfiltration and unintended data exposure

System Integrity Threats
Unauthorized modification and configuration exploitation

Novel Attack Vectors
Prompt injection and model extraction attacks

Evaluation methodology components

Threat modeling approaches involve systematically analyzing potential attack paths and assessing the capability and motivation of potential attackers.

This also includes evaluating the effectiveness of defense-in-depth strategies, identifying trust boundaries within the system, and assessing the adequacy of existing security controls.

Vulnerability assessment techniques: rely on structured security testing methodologies. These include reviewing system configurations, analyzing code and model security, evaluating third-party components, and verifying operational security to uncover weaknesses.

Impact estimation frameworks help determine the potential consequences of security incidents.

This includes analyzing the impact of confidentiality breaches, assessing the effects of integrity violations, and evaluating how disruptions to availability could affect operations. Reputational damage and potential financial losses are also key considerations in understanding the full scope of impact.

Organizations with sophisticated security risk evaluation frameworks typically identify and address vulnerabilities before they can be exploited, creating digital workforces that maintain security even as threat landscapes evolve.

Operational Risk Considerations

Beyond security concerns, digital workers introduce operational risks that impact business continuity, performance, and reliability. These risks require distinct evaluation approaches.

Operational risk categories

1. **Availability and reliability risks**

- System failure scenarios
- Dependency availability concerns

- Performance degradation pathways
- Capacity limitation impacts
- Recovery capability gaps

2. **Process integration risks**

- Workflow disruption possibilities
- Handoff failure scenarios
- Exception handling inadequacies
- Business rule misalignment
- Process boundary ambiguities

3. **Quality and performance risks**

- Accuracy degradation scenarios
- Consistency variation concerns
- Edge case handling limitations
- Drift and degradation possibilities
- Knowledge currency challenges

4. **Vendor and provider risks**

- Service interruption scenarios
- Contractual protection gaps
- Provider viability concerns
- Version compatibility issues
- Support quality uncertainties

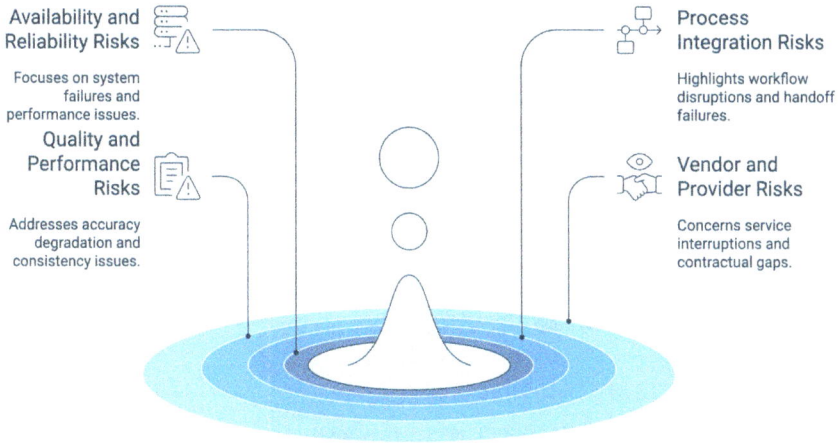

Operational Risk Categories

Availability and Reliability Risks
Focuses on system failures and performance issues.

Quality and Performance Risks
Addresses accuracy degradation and consistency issues.

Process Integration Risks
Highlights workflow disruptions and handoff failures.

Vendor and Provider Risks
Concerns service interruptions and contractual gaps.

Assessment approach elements

Failure mode analysis focuses on systematically identifying potential points of failure within a system. It includes evaluating the chain of consequences that could result from each failure, assessing the effectiveness of detection mechanisms, reviewing recovery approaches, and analyzing the system's ability to prevent failures.

Dependency mapping involves creating a comprehensive inventory of all dependencies. Each dependency is assessed for its criticality, availability of alternatives, and associated concentration risks. Based on these findings, mitigation strategies are developed to ensure resilience.

Business impact analysis: examines the potential consequences of disruptions. It includes determining the maximum tolerable downtime, setting recovery time objectives, quantifying financial impact, assessing operational disruptions, and evaluating potential effects on organizational reputation.

A mature operational risk assessment framework helps digital workforces develop appropriate resilience, reliability, and performance characteristics for their intended business functions.

Reputational Risk Management

The deployment of digital workers, particularly those with customer-facing responsibilities, introduces significant reputational risk that requires dedicated assessment and management approaches.

Reputational risk sources

1. **Customer experience impacts**

- Inappropriate response scenarios
- Empathy and understanding limitations
- Accessibility barrier possibilities
- Personalization inadequacies
- Service consistency challenges

2. **Ethical concern triggers**

- Bias manifestation possibilities
- Fairness perception issues
- Transparency limitation concerns
- Agency and autonomy questions
- Value alignment challenges

3. **Public perception factors**

- Employment impact narratives
- Technology anxiety triggers
- Privacy concern activation
- Control and oversight questions
- Authenticity and disclosure expectations

4. Incident amplification mechanisms

- Social media escalation pathways
- Traditional media coverage patterns
- Stakeholder complaint channels
- Competitive exploitation possibilities
- Regulatory attention triggers

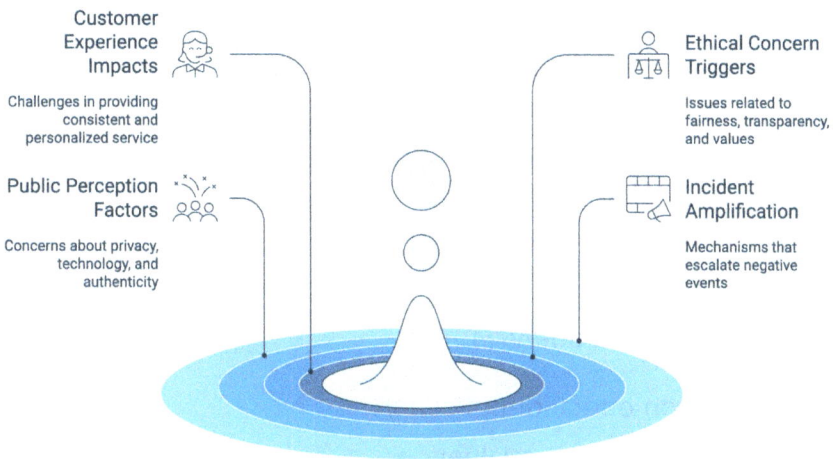

Reputational Risk Sources

Customer Experience Impacts
Challenges in providing consistent and personalized service

Public Perception Factors
Concerns about privacy, technology, and authenticity

Ethical Concern Triggers
Issues related to fairness, transparency, and values

Incident Amplification
Mechanisms that escalate negative events

Assessment methodologies

Perception analysis approaches focus on understanding how digital systems are viewed by stakeholders. This includes mapping stakeholder expectations, monitoring sentiment, identifying recurring themes of concern, evaluating the effectiveness of communication, and tracking how the narrative evolves over time.

Scenario planning techniques help organizations prepare for uncertainties by developing "what if" scenarios, mapping out possible consequence chains, preparing response strategies, identifying early warning indicators, and planning mitigation approaches in advance.

Experience testing programs are designed to evaluate system performance and usability. They include testing with diverse user interactions, analyzing responses to edge cases, applying progressive stress testing, observing long-term interaction patterns, and conducting comparative assessments to benchmark user experience.

Reputational risk requires particular attention because digital workers, as highly visible manifestations of organizational values and priorities, can rapidly impact brand perception and stakeholder trust if not carefully managed.

Integrated Risk Assessment Approaches

The most effective risk management programs integrate these various risk domains into comprehensive assessment approaches that capture how risks interact with each other and enable holistic prioritization. Integration methodology elements include:

1. **Cross-domain analysis**

- Interaction effect identification
- Cascading impact evaluation
- Common root cause discovery
- Mitigation synergy exploration
- Conflicting control recognition

2. **Unified evaluation frameworks**

- Standardized assessment criteria
- Comparable impact measurement
- Consistent likelihood estimation
- Holistic risk visualization
- Integrated prioritization mechanisms

3. **Lifecycle integration**

- Design phase risk assessment
- Development stage evaluation
- Pre-deployment verification
- Operational monitoring
- Enhancement and evolution assessment

Building a Holistic Risk Strategy

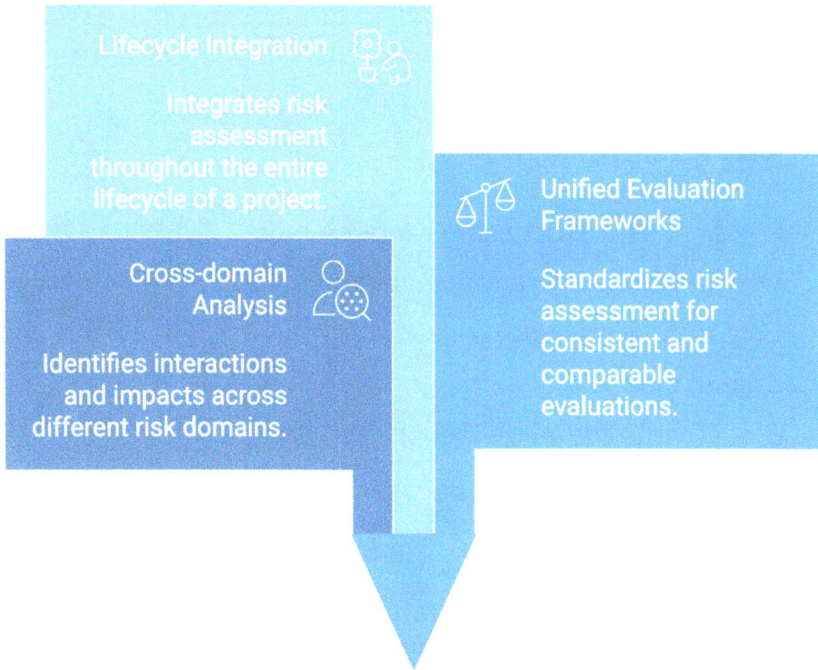

Lifecycle Integration

Integrates risk assessment throughout the entire lifecycle of a project.

Cross-domain Analysis

Identifies interactions and impacts across different risk domains.

Unified Evaluation Frameworks

Standardizes risk assessment for consistent and comparable evaluations.

Organizations with mature risk management recognize that digital workforce risks cannot be effectively managed in isolation. It requires integrated approaches that address the complex interplay between different risk domains.

Regulatory Compliance: Navigating Evolving Legal Requirements

Digital workers operate in an increasingly complex regulatory environment with evolving requirements across multiple domains. Organizations must develop systematic approaches to understanding obligations and ensuring compliance.

Current Regulatory Landscape Overview

The regulatory environment for digital workers spans multiple domains, with significant variation by geography, industry, and application context. Understanding this landscape is essential for effective compliance management.

Key regulatory domains

1. **Data protection and privacy**

- Personal data processing requirements
- Consent and transparency obligations
- Cross-border transfer restrictions
- Data subject rights frameworks
- Breach notification obligations

2. **Algorithmic governance**

- Fairness and non-discrimination requirements
- Explainability and transparency obligations
- Human oversight provisions
- Impact assessment mandates
- Testing and documentation requirements

3. **Industry-specific regulation**

- Financial services obligations
- Healthcare information requirements

- Critical infrastructure regulations
- Professional services standards
- Consumer protection provisions

4. **General business regulation**

- Labor and employment standards
- Intellectual property protections
- Contract and commercial law
- Trade and competition provisions
- Corporate governance requirements

Geographic variation considerations

1. **Jurisdictional differences**

- European Union regulatory approaches
- North American frameworks
- Asia-Pacific requirements
- Global variation in enforcement intensity
- Cross-border compliance complexity

2. **Emerging regulatory trends**

- Increasing AI-specific legislation
- Sector-based regulatory expansion
- International standards development
- Self-regulatory framework evolution
- Enforcement priority shifts

Organizations must develop mechanisms to monitor this evolving landscape, interpret requirements relevant to their specific implementations, and translate these into operational compliance controls and processes.

Compliance Strategies and Documentation

Effective compliance requires systematic approaches to understanding, implementing, and demonstrating adherence to regulatory requirements.

Strategy development elements

1. **Obligation identification**

- Comprehensive regulatory mapping
- Applicability assessment for specific implementations
- Requirement interpretation for operational context
- Materiality determination for prioritization
- Update monitoring for evolving obligations

2. **Control implementation**

- Control design for requirement fulfillment
- Implementation planning and execution
- Effectiveness testing methodologies
- Remediation management for gaps
- Ongoing verification processes

3. **Documentation frameworks**

- Purpose and process documentation
- Design and development records
- Testing and validation evidence
- Operational monitoring logs
- Incident and response recording

4. **Governance and oversight**

- Responsibility assignment
- Review and approval processes

- Regular assessment cadence
- Escalation and exception handling
- Executive awareness and accountability

Strategy Development Elements

Obligation Identification	Control Implementation	Documentation Frameworks	Governance and Oversight

Implementation approach options

Integrated compliance frameworks aim to streamline governance by establishing a unified structure. This includes developing a common control catalog, applying a consistent assessment methodology, coordinating reporting mechanisms, and maintaining a centralized repository for documentation.

Risk-based prioritization focuses on aligning implementation efforts with potential impact. It involves assessing the materiality of requirements, evaluating the consequences of violations, considering implementation complexity, optimizing resource allocation, and adopting a progressive enhancement approach.

External assurance integration brings in independent perspectives to validate compliance. This includes incorporating third-party assessments, pursuing certifications where they add value, preparing for audits, engaging proactively with regulators, and aligning with relevant industry standards.

Effective compliance strategies balance the need for comprehensive obligation fulfillment with practical implementation approaches that minimize unnecessary burden while providing appropriate protection.

Anticipating Future Regulatory Developments

The regulatory landscape for AI and digital workers continues to evolve rapidly. Forward-thinking organizations develop approaches to anticipate likely developments and prepare proactively rather than reacting to each new requirement.

Anticipation methodology components

1. **Trend monitoring**

- Legislative proposal tracking
- Regulatory guidance analysis
- Enforcement action evaluation
- Industry standard evolution
- Public opinion and advocacy trends

2. **Impact assessment**

- Digital workforce implications
- Implementation requirement analysis
- Timeline and resource estimation
- Strategic response option development
- Competitive positioning consideration

3. **Proactive preparation**

- "Future-proofing" design approaches
- Documentation enhancement for anticipated needs
- Control framework evolution planning
- Capability building for likely requirements
- Strategic influence and engagement

Key trend areas

1. **Transparency and explainability**

- Increasing requirements for understandable AI
- Documentation obligations for decision processes
- User notification about AI interaction
- Evidence preservation for review
- Meaningful explanation standards

2. **Human oversight and responsibility**

- Expanding human-in-the-loop requirements
- Accountability framework definition
- Intervention capability mandates
- Supervision standard development
- Liability framework evolution

3. **Testing and validation**

- Pre-deployment assessment requirements
- Ongoing monitoring obligations
- Performance standard development
- Bias and fairness testing mandates
- Safety and reliability verification

AI Governance Trends

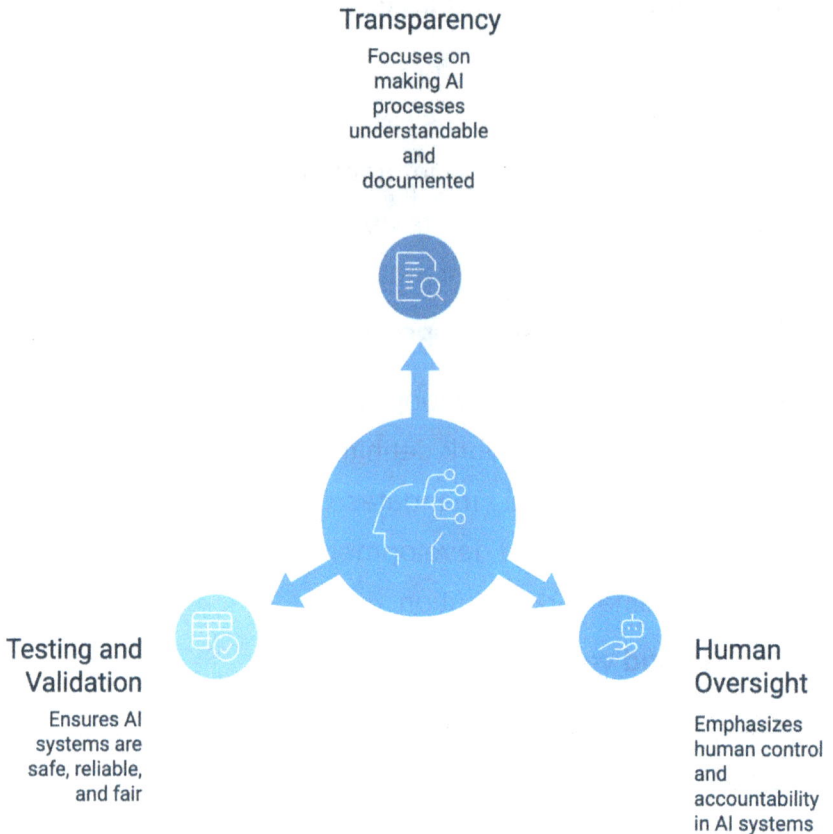

Transparency

Focuses on making AI processes understandable and documented

Testing and Validation

Ensures AI systems are safe, reliable, and fair

Human Oversight

Emphasizes human control and accountability in AI systems

Organizations that anticipate regulatory evolution design digital workers that not only meet current requirements but also adapt efficiently to future obligations without requiring redesign.

Ethical Frameworks: Ensuring Responsible Deployment

Beyond regulatory compliance, organizations must address broader ethical considerations to ensure digital workers operate responsibly and maintain stakeholder trust. This requires systematic approaches to identifying and addressing ethical dimensions.

Bias Mitigation Strategies

Digital workers can perpetuate or amplify existing biases, requiring dedicated strategies to identify, measure, and mitigate these effects throughout the lifecycle.

Bias source categories

1. **Data-driven bias**

- Training data representation imbalance
- Historical pattern perpetuation
- Measurement and label bias
- Sampling methodology limitations
- Temporal drift effects

2. **Design-induced bias**

- Problem framing assumptions
- Feature selection decisions
- Optimization objective choices
- Parameter weighting approaches
- Testing limitation effects

3. **Deployment context bias**

- Access inequality implications
- Usage pattern differences
- Feedback loop effects
- Interpretation variation
- Environment adaptation challenges

Bias in AI systems stems from various hidden sources.

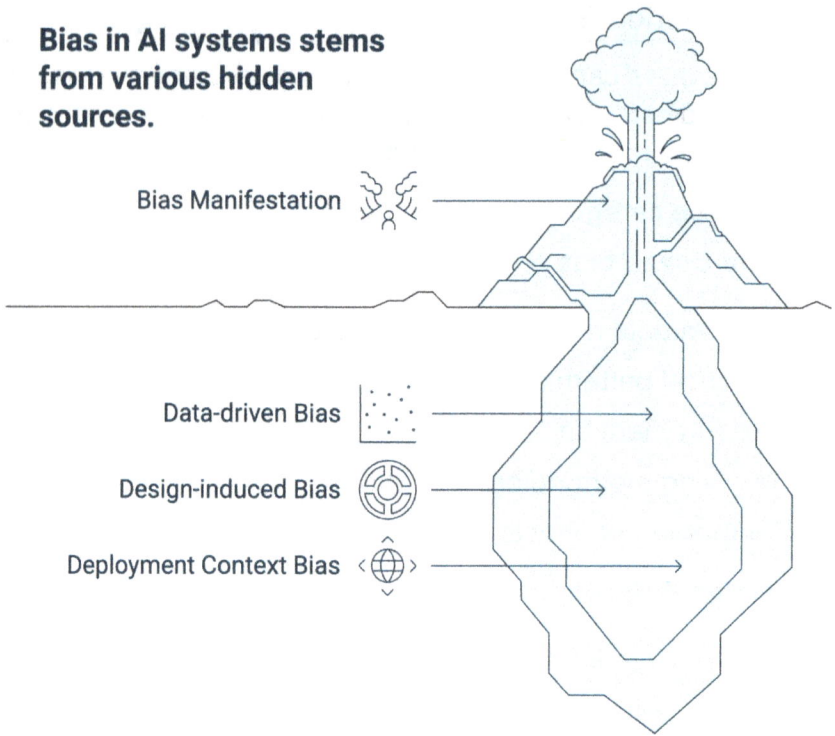

Bias Manifestation

Data-driven Bias

Design-induced Bias

Deployment Context Bias

Mitigation approach elements

1. Comprehensive assessment

- Multiple bias metric utilization
- Disaggregated analysis by subgroup
- Intersectional evaluation
- Contextual impact consideration
- Ongoing monitoring rather than one-time testing

2. Technical interventions

- Data augmentation and balancing
- Fairness constraint implementation
- Model selection and tuning approaches
- Pre-processing and post-processing techniques

- Ensemble and verification methods

3. **Process enhancements**

- Diverse development team formation
- External review incorporation
- Stakeholder input integration
- Documentation and transparency practices
- Continuous improvement processes

The most effective bias mitigation approaches recognize that addressing bias requires ongoing attention throughout the digital worker lifecycle rather than one-time solutions at a single development stage.

Fairness Assessment Methodologies

Fairness is a multidimensional concept that extends beyond bias measurement to broader questions of equitable impact, appropriate consistency, and just outcomes.

Fairness dimension framework

1. **Distributive justice**

- Outcome equity across groups
- Resource allocation appropriateness
- Opportunity access equality
- Benefit and burden distribution
- Comparative advantage consideration

2. **Procedural justice**

- Process consistency and standardization
- Voice and participation opportunities
- Transparency and understandability

- Appeal and recourse availability
- Decision criteria appropriateness

3. **Representational justice**

- Respectful and dignified treatment
- Cultural sensitivity and appropriateness
- Stereotype avoidance
- Individual recognition
- Agency preservation

Understanding Fairness Dimensions

Representational
Justice

Highlights respectful
treatment and cultural
sensitivity

Distributive
Justice

Focuses on equitable
outcomes and resource
allocation

Procedural
Justice

Emphasizes fair
processes and
transparency

Assessment methodology components

1. **Quantitative measurement**

- Group parity metrics
- Calibration analysis
- Error rate comparison

- Benefit and harm distribution
- Longitudinal impact tracking

2. **Qualitative evaluation**

- Stakeholder perspective gathering
- Experience analysis across groups
- Contextual appropriateness assessment
- Value alignment verification
- Edge case and outlier consideration

3. **Participatory approaches**

- Affected community involvement
- Diverse perspective integration
- Lived experience incorporation
- Stakeholder feedback mechanisms
- Collaborative definition development

Organizations that implement advanced fairness assessment frameworks are better equipped to create digital workers that appropriately balance multiple fairness dimensions. This approach prevents them from focusing on a single fairness metric, which may create unintended consequences.

Transparency and Explainability Approaches

Digital workers, particularly those based on complex machine learning models, can appear opaque in their decision-making. Effective implementation requires appropriate transparency and explainability to build trust and enable appropriate oversight.

Transparency level framework

1. **System transparency**

- Purpose and capability disclosure
- Limitation and boundary communication
- Data usage and handling explanation
- Human role and oversight description
- Performance metric publication

2. **Process transparency**

- Decision criteria explanation
- Information source disclosure
- Confidence level indication
- Alternative option presentation
- Reasoning process description

3. **Outcome transparency**

- Result explanation provision
- Factor influence communication
- Counterfactual possibility description
- Comparable case reference
- Improvement pathway indication

Achieving Transparency in AI Systems

System Transparency
Disclosing purpose, capabilities, and limitations

Process Transparency
Explaining decision criteria and reasoning

Outcome Transparency
Providing result explanations and improvement paths

Implementation methodology elements

Audience-appropriate approaches focus on customizing explanations to different stakeholder needs. This involves calibrating technical depth, optimizing the format for clarity, ensuring sensitivity to timing and context, and using progressive disclosure to present information in manageable layers.

Technical enablement emphasizes methods that support transparency and interpretability. This includes selecting interpretable models where feasible, integrating post-hoc explanation techniques, conducting feature importance analysis, extracting decision trees, and generating example-based explanations to illustrate system behavior.

Process enhancement supports explanation quality and traceability. Key elements include maintaining thorough documentation, implementing mechanisms for recording decisions, enabling review capabilities, verifying the quality of explanations, and incorporating feedback to drive continuous improvement.

The most effective transparency approaches recognize that explanation needs vary significantly by context, audience, and decision type, requiring flexible frameworks rather than one-size-fits-all solutions.

Responsible Innovation Governance

Organizations require comprehensive governance frameworks that ensure responsible innovation throughout the digital worker lifecycle.

Governance framework components

1. **Value identification and prioritization**

- Organizational value articulation
- Stakeholder value consideration
- Societal impact reflection
- Trade-off guidance development

- Value evolution mechanisms

2. **Ethical risk assessment**

- Systematic impact evaluation
- Vulnerable population consideration
- Unintended consequence exploration
- Misuse potential analysis
- Long-term implication examination

3. **Stakeholder engagement**

- Affected community consultation
- Expert input incorporation
- Cross-functional perspective integration
- Public feedback mechanisms
- Ongoing dialogue maintenance

4. **Review and oversight mechanisms**

- Ethics review process implementation
- Stage-gate assessment approaches
- Independent evaluation incorporation
- Continuous monitoring frameworks
- Post-deployment audit processes

Governance Framework Structure

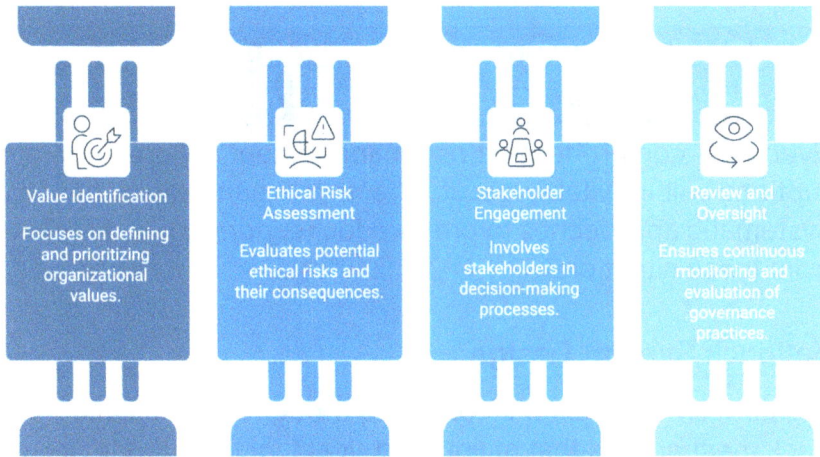

Value Identification	Ethical Risk Assessment	Stakeholder Engagement	Review and Oversight
Focuses on defining and prioritizing organizational values.	Evaluates potential ethical risks and their consequences.	Involves stakeholders in decision-making processes.	Ensures continuous monitoring and evaluation of governance practices.

Implementation approaches

Dedicated ethics functions focus on formal structures that guide ethical decision-making. This includes establishing ethics committees, creating dedicated ethics officer roles, engaging external advisors, integrating ethics specialists into teams, and fostering a broader ethics community within the organization.

Integrated development processes embed ethics into the technical workflow. This is achieved through ethics-by-design methodologies, requiring impact assessments, enforcing documentation and review practices, setting testing and validation standards, and establishing mechanisms for continuous improvement.

Building a cultural foundation involves demonstrating leadership commitment to ethics, developing awareness across teams, encouraging open discussion of values, ensuring psychological safety for raising concerns, and recognizing individuals who actively consider ethical implications.

Organizations with mature ethics governance typically create digital workforces that not only avoid harm but actively create positive value aligned with organizational and societal priorities.

Even with comprehensive risk management, failures and disruptions will inevitably occur. Effective organizations develop systematic approaches to prepare for, respond to, and recover from these situations.

Fallback System Design

Digital workers should incorporate thoughtfully designed fallback mechanisms that maintain critical functions if primary capabilities are compromised.

Fallback design principles

1. **Graceful degradation**

- Progressive capability reduction
- Critical function preservation
- Performance level adjustment
- Resource reallocation under constraints
- User experience adaptation

2. **Failure isolation**

- Component independence protection
- Cascading failure prevention
- Fault containment mechanisms
- Clean boundary maintenance
- Partial operation capability

3. **Redundancy and diversity**

- Critical component duplication
- Alternative approach implementation
- Different technology utilization
- Independent resource allocation
- Geographic distribution, where relevant

Fallback Design Principles

Redundancy and Diversity
Ensuring system resilience through duplication and alternative approaches

Graceful Degradation
Progressive reduction of capabilities while preserving critical functions

Failure Isolation
Preventing cascading failures by isolating components

Implementation patterns

Tiered capability frameworks structure system functionality in layers. This includes identifying core functions, building enhanced capabilities on top, and defining progressive fallback sequences.

Establishing a minimum viable service helps maintain operations under constraint, while setting recovery priorities ensures essential services are restored first.

Alternative path design ensures continuity through backup mechanisms. These include secondary processing options, manual override capabilities, simplified fallback approaches, rule-based alternatives, and the preservation of legacy systems for critical functions.

Hybrid human-digital approaches integrate human involvement where needed. This includes defining escalation paths to human operators, adjusting levels of supervision, enabling collaborative recovery processes, establishing protocols for responsibility transfer, and shifting communication methods based on the situation.

Effective fallback design creates digital workers that maintain reliability and continuity even when facing unexpected challenges or limitations.

Business Continuity Planning

Beyond technical fallbacks, organizations must develop comprehensive business continuity plans that address the broader operational implications of digital worker disruption.

Continuity planning elements

1. **Impact assessment**

- Critical process identification
- Dependency mapping
- Recovery time objective determination
- Operational impact quantification
- Financial consequence estimation

2. **Continuity strategy development**

- Alternative process design

- Resource requirement identification
- Responsibility assignment
- Trigger condition definition
- Communication approach planning

3. **Implementation and preparation**

- Documentation development
- Resource allocation and reservation
- Training and awareness building
- Testing and validation
- Maintenance and updating processes

Continuity Planning Cycle

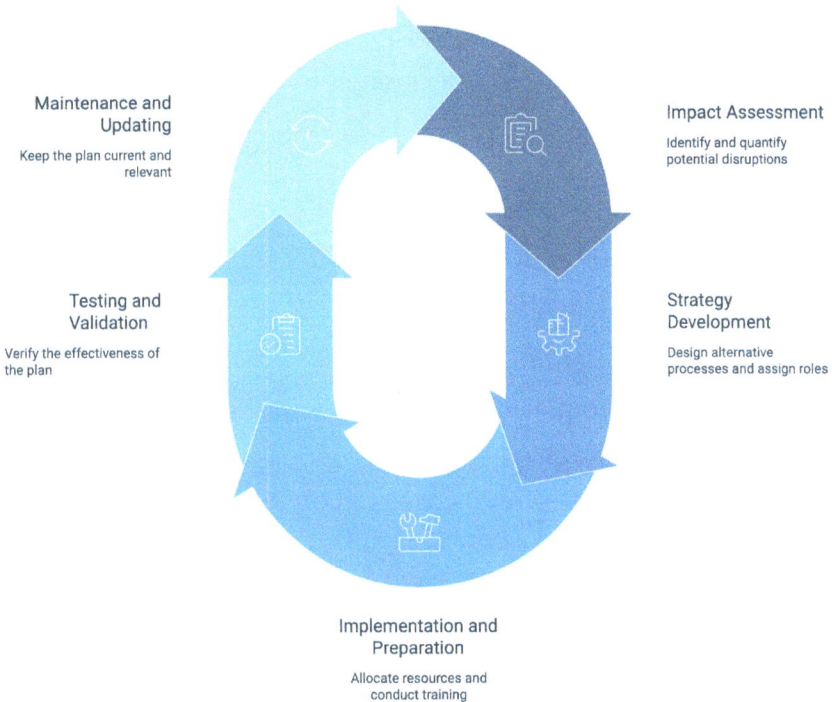

Maintenance and Updating
Keep the plan current and relevant

Impact Assessment
Identify and quantify potential disruptions

Testing and Validation
Verify the effectiveness of the plan

Strategy Development
Design alternative processes and assign roles

Implementation and Preparation
Allocate resources and conduct training

Strategy option categories

1. **Technical resilience approaches**

- Redundant infrastructure implementation
- Geographic distribution planning
- Backup system maintenance
- Recovery automation
- Alternative provider arrangements

2. **Process adaptation strategies**

- Manual workaround development
- Temporary process simplification
- Service level adjustment planning
- Customer communication approaches
- Priority-based recovery sequencing

3. **Resource flexibility mechanisms**

- Cross-training for recovery roles
- External resource pre-arrangement
- Surge capacity maintenance
- Asset reallocation frameworks
- Emergency procurement processes

Organizations with sophisticated continuity planning typically maintain operational effectiveness even during significant digital worker disruptions, minimizing business impact while implementing recovery solutions.

Recovery Procedures and Protocols

When incidents occur despite preventive measures, effective recovery requires clearly defined procedures and protocols that enable rapid restoration of normal operations.

Recovery framework components

1. **Incident response processes**

- Detection and declaration mechanisms
- Initial assessment procedures
- Containment approach options
- Escalation and notification protocols
- Role and responsibility activation

2. **Recovery execution**

- Structured recovery procedure activation
- Priority-based restoration sequencing
- Progress tracking mechanisms
- Effectiveness verification processes
- Adjustment approach for challenges

3. **Return to normal operations**

- Transition criteria and verification
- Phased resumption planning
- Performance monitoring enhancement
- User communication approaches
- Post-incident support provision

Incident Recovery Process Funnel

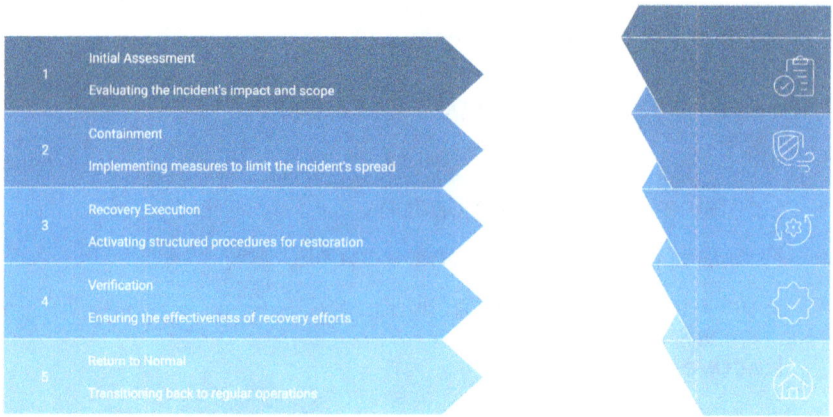

	Stage	Description
1	Initial Assessment	Evaluating the incident's impact and scope
2	Containment	Implementing measures to limit the incident's spread
3	Recovery Execution	Activating structured procedures for restoration
4	Verification	Ensuring the effectiveness of recovery efforts
5	Return to Normal	Transitioning back to regular operations

Protocol design considerations

Documentation development focuses on creating clear and accessible protocols. This involves providing scenario-specific guidance, incorporating decision support tools, including relevant contact and resource information, and ensuring clarity around authority and approval structures.

Testing and validation ensure that protocols are practical and effective. Methods include conducting tabletop exercises, running simulation-based tests, performing full-scale recovery rehearsals, and integrating learnings from each test into continuous improvement processes.

Organizational integration embeds protocols into operations. This includes launching awareness and training initiatives, clearly defining roles and responsibilities, regularly reviewing and updating protocols, engaging leadership, and promoting coordination across functional teams.

The most effective recovery approaches combine technical solutions with organizational readiness, creating the capabilities necessary to manage disruptions effectively when they inevitably occur.

Learning from Incidents

Organizations should implement systematic approaches to learn from incidents and enhance future resilience based on these experiences.

Learning system elements

1. **Incident analysis**

- Root cause investigation
- Contributing factor identification
- Timeline and sequence documentation
- Impact assessment
- Response effectiveness evaluation

2. **Improvement identification**

- Prevention opportunity recognition
- Detection enhancement possibilities
- Response improvement options
- Recovery acceleration approaches
- Resilience strengthening strategies

3. **Knowledge integration**

- Lesson documentation and sharing
- Policy and procedure updating
- Training and awareness enhancement
- Design principle evolution
- Architecture and implementation adjustment

Learning System Cycle

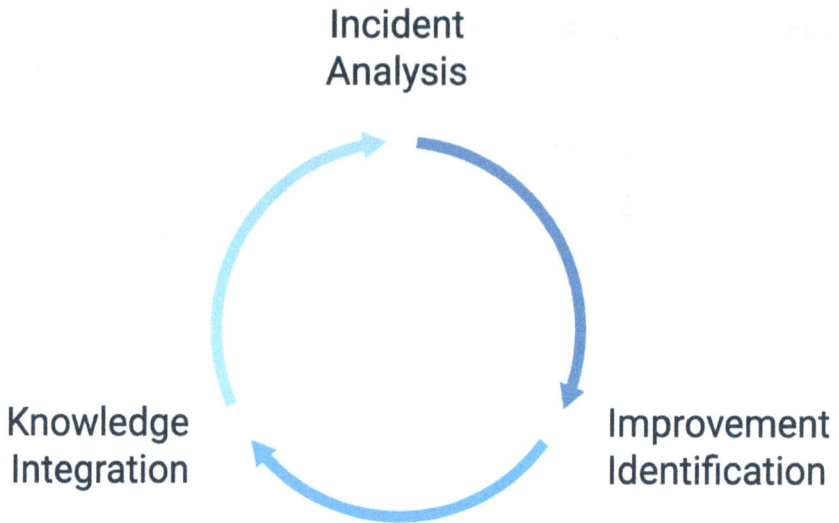

Incident
Analysis

Knowledge
Integration

Improvement
Identification

Implementation approaches

Formal post-incident review involves conducting structured analyses following incidents. These reviews include participation from cross-functional teams, fostering a blameless investigation culture, maintaining a systemic perspective, and developing actionable plans with clear tracking for follow-through.

Pattern recognition across incidents focuses on learning from historical data. This includes identifying common causes, analyzing trends over time, recognizing recurring vulnerability patterns, comparing the effectiveness of past responses, and evaluating different recovery approaches.

Preventive enhancement programs aim to strengthen systems by applying lessons learned. These programs involve the systematic implementation of improvements, planning enhancements based on priority, allocating resources to build resilience, verifying the effectiveness of changes, and maintaining a continuous learning cycle to support long-term improvement.

Advanced incident learning capabilities provide organizations with increased resilience over time as they incorporate lessons from each experience into enhanced prevention, detection, and recovery capabilities.

Conclusion: Balancing Innovation and Protection

Effective risk management and compliance for digital workers requires balancing two seemingly contradictory imperatives: enabling innovation and transformation while providing appropriate protections against potential harm.

Organizations that excel in this domain develop approaches that address both needs simultaneously rather than treating them as trade-offs.

Several key principles distinguish successful approaches:

1. **Risk-aware innovation**: Considering associated risks throughout the innovation process rather than treating it as a separate compliance checkpoint enables early identification and mitigation of potential issues.

2. **Proportional control**: Calibrating protection measures to the specific risk profile of each implementation rather than applying one-size-fits-all approaches that may unnecessarily constrain low-risk applications.

3. **Stakeholder inclusion**: Engaging diverse perspectives in risk identification and management to ensure a comprehensive understanding of potential impacts across different groups and contexts.

4. **Continuous evolution**: Adopting adaptive approaches that recognize both risks and requirements change over time, rather than relying on static controls established once and left unchanged.

5. **Positive value creation**: Moving beyond harm prevention to actively designing for positive impact aligned with organizational values and societal benefit.

As organizations expand the roles and responsibilities of digital workers, those that implement sophisticated risk management and compliance approaches will be better positioned to capture the full potential of these technologies while maintaining stakeholder trust and regulatory compliance.

CHAPTER 9:

GOVERNANCE FOR THE DIGITAL ERA

TL;DR:

- Effective digital workforce governance requires clear accountability frameworks with well-defined roles, decision rights, and documentation requirements that establish responsibility while enabling appropriate autonomy.

- Thoughtful policy development creates guardrails for implementation through usage frameworks, boundary conditions, and exception handling procedures that evolve as capabilities and needs change.

- Cross-functional oversight structures, including appropriately designed committees, review processes, escalation pathways, and coordination mechanisms, enable consistent governance across organizational boundaries.

- External stakeholder management extends governance beyond organizational boundaries to address customer expectations, partner relationships, regulatory engagement, and broader societal responsibilities.

- Successful organizations view governance as an enabler of sustainable value rather than merely a constraint, implementing balanced approaches that evolve with changing capabilities and needs.

Effective Workforce Governance

As digital workers become increasingly embedded in organizational operations, traditional governance approaches prove inadequate for managing their unique characteristics and implications.

The autonomous, evolving, and often opaque nature of agentic AI systems requires new governance frameworks that balance innovation with appropriate control.

Accountability Frameworks: Establishing Clear Lines of Responsibility

A fundamental challenge in digital workforce governance is determining who is responsible for what as autonomous systems operate with limited direct human supervision. Effective governance requires clarity on these questions.

Organizational Roles and Responsibilities

Digital worker governance necessitates a clear definition of roles and responsibilities across multiple organizational functions and levels.

Key governance stakeholder categories

1. **Executive leadership**

- Strategic direction setting
- Resource allocation approval
- Organizational alignment facilitation
- Accountability for overall outcomes
- Risk tolerance definition and communication

2. **Digital workforce program leadership**

- Implementation strategy development
- Cross-functional coordination

- Standard and methodology establishment
- Capability building orchestration
- Performance monitoring and reporting

3. **Business function owners**

- Use case identification and prioritization
- Process integration responsibility
- Value realization accountability
- Change management leadership
- Operational oversight of deployed systems

4. **Technical implementation teams**

- Design and development execution
- Technical standard adherence
- Integration with existing systems
- Performance optimization
- Technical maintenance and enhancement

5. **Risk and compliance functions**

- Control framework development
- Risk assessment facilitation
- Compliance verification
- Policy interpretation and guidance
- Audit coordination and response

Responsibility definition approaches

RACI frameworks clarify roles by categorizing responsibilities as Responsible (those who do the work), Accountable (those ultimately answerable), Consulted (those who provide input), and Informed (those kept updated). This structure ensures clear mapping of responsibilities across different stages of the lifecycle.

Stage-gate authority models define specific decision points throughout a project or process lifecycle. Each gate includes clear approval requirements, calibrated authority levels, specified escalation paths, and mechanisms for documentation and verification.

Decision rights assignment involves explicitly delegating specific authorities, defining boundary conditions, and outlining oversight requirements. It includes setting a cadence for review and reassessment, ensuring a balance between individual autonomy and necessary control.

Responsibility definition approaches range from broad to specific.

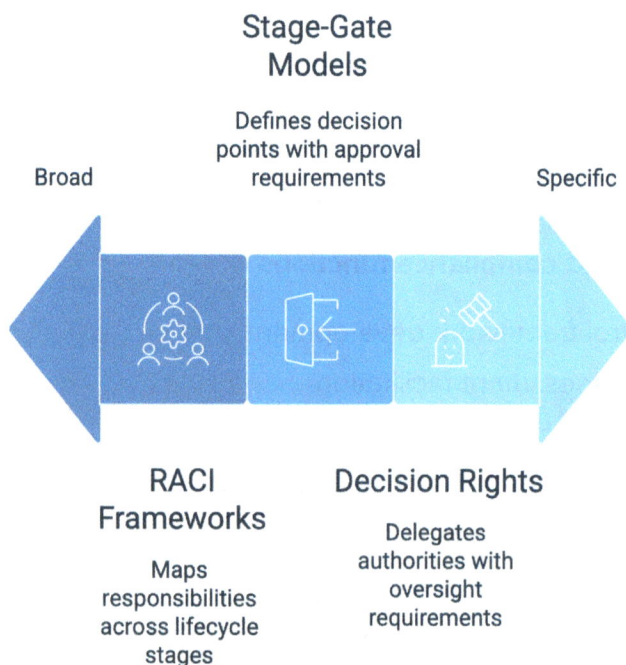

Stage-Gate
Models

Defines decision
points with approval
requirements

Broad   Specific

RACI
Frameworks

Maps
responsibilities
across lifecycle
stages

Decision Rights

Delegates
authorities with
oversight
requirements

Organizations with well-defined responsibility frameworks typically experience fewer governance failures, more efficient decision-making, and clearer paths for issue resolution when problems inevitably arise.

Decision Rights Allocation

Beyond general role definition, effective governance requires explicit determination of who can make which decisions regarding digital workers throughout their lifecycle.

Decision domain categories

1. **Strategic decisions**

- Investment prioritization
- Risk tolerance determination
- Organizational impact management
- Strategic partnership formation
- Long-term direction setting

2. **Implementation decisions**

- Use case selection and sequencing
- Technology and architecture choices
- Resource allocation and timing
- Methodology and approach selection
- Integration strategy determination

3. **Operational decisions**

- Performance parameter setting
- Exception handling protocols
- Change and enhancement approval
- Interaction model configuration
- Monitoring approach definition

4. **Risk and compliance decisions**

- Control design and implementation
- Audit scope and frequency
- Issue remediation prioritization
- Policy exception management
- Incident response direction

Which decision domain should be prioritized?

Strategic Decisions
Focus on long-term direction and investment prioritization.

Implementation Decisions
Concentrate on technology choices and resource allocation.

Operational Decisions
Manage performance parameters and exception handling.

Risk and Compliance Decisions
Ensure control design and incident response.

Allocation approach elements

Centralization versus decentralization calibration determines the appropriate balance between enterprise-level and business unit authority.

This requires managing trade-offs between standardization and customization, weighing efficiency against local alignment, and considering the benefits of scale versus the need for responsiveness. The appropriate level of decision-making authority is determined based on the nature of each decision.

Establishing the authority threshold focuses on assigning decision levels based on impact. It considers risk exposure, evaluates precedent and novelty, assesses resource requirements, and weighs strategic importance to determine the appropriate level of authority.

Delegation and escalation mechanisms ensure clarity in how decisions are transferred or escalated. This includes defining processes for authority transfer, setting thresholds for higher-level review, outlining exception handling and emergency decision protocols, and providing for temporary authority when needed.

The most effective decision rights frameworks balance the need for appropriate control with the flexibility required for innovation and responsiveness. This calibrates authority levels to the specific context and consequences of different decision types.

Audit Trails and Documentation Requirements

Accountability for digital workers requires comprehensive mechanisms to record decisions, actions, and outcomes throughout the lifecycle, creating transparency and enabling verification.

Documentation domain framework

1. **Design and development documentation**

- Requirements and specifications
- Design decisions and rationales
- Testing approaches and results
- Performance validation evidence
- Risk assessment and mitigation planning

2. **Operational documentation**

- Configuration parameters and changes
- Performance monitoring records
- Exception handling instances
- Enhancement implementation details
- Incident occurrence and resolution

3. **Governance process documentation**

- Approval decisions and justifications
- Review findings and actions

- Policy exception management
- Audit activities and outcomes
- Issue remediation tracking

4. **Value and impact documentation**

- Performance against objectives
- Business impact measurement
- User satisfaction assessment
- Problem resolution effectiveness
- Continuous improvement activities

Documentation Domain Framework

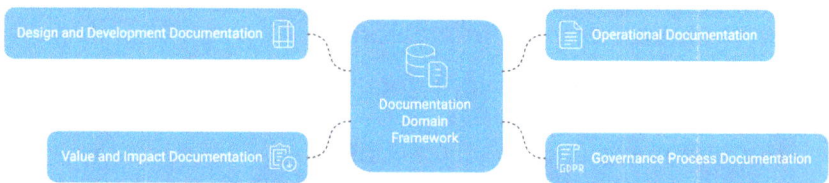

Design and Development Documentation

Operational Documentation

Documentation Domain Framework

Value and Impact Documentation

Governance Process Documentation

Implementation approach elements

1. **Documentation system design**

- Appropriate detail level determination
- Access and searchability optimization
- Integration with existing repositories
- Automation where feasible
- Usability focus for actual utilization

2. **Responsibility assignment**

- Clear ownership of documentation domains
- Verification and quality control processes
- Completeness assessment mechanisms
- Currency maintenance requirements

- Cross-functional coordination

3. **Proportional approach application**

- Risk-based documentation requirements
- Impact-appropriate detail levels
- Value-driven prioritization
- Burden minimization focus
- Continuous refinement based on utility

The most effective documentation frameworks balance the need for comprehensive records with practical considerations of maintenance burden. They prioritize high-value documentation that enables actual governance rather than creating compliance artifacts with limited utility.

Policy Development: Creating Guidelines for Digital Workforce Deployment

Effective governance requires clear policies that establish boundaries, standards, and expectations for digital worker implementation and operation across the organization.

Usage Policy Frameworks

Usage policies define how, when, and for what purposes digital workers may be employed within the organization, establishing guardrails for appropriate application.

Policy domain categories

1. **Appropriate use determination**

- Authorized application categories
- Prohibited or restricted uses
- Approval requirements by use type
- User qualification standards

- Purpose limitation principles

2. **Access and authentication**

- User authorization requirements
- Authentication mechanism standards
- Delegation and impersonation rules
- Session management expectations
- Credential protection obligations

3. **Data handling guidelines**

- Permitted data source specification
- Sensitive data protection requirements
- Retention and deletion standards
- Data quality expectations
- Information sharing limitations

4. **Performance expectations**

- Quality and accuracy standards
- Response time expectations
- Availability requirements
- Monitoring and reporting obligations
- Improvement process expectations

Policy Domain Categories

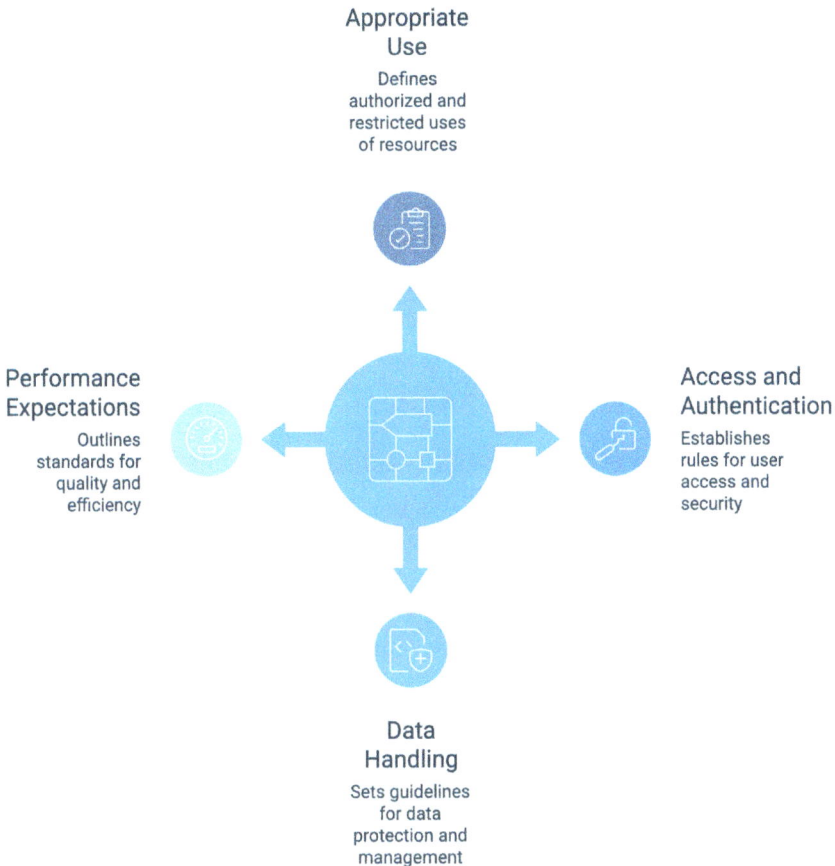

Appropriate Use

Defines authorized and restricted uses of resources

Performance Expectations

Outlines standards for quality and efficiency

Access and Authentication

Establishes rules for user access and security

Data Handling

Sets guidelines for data protection and management

Policy development approaches

Risk-based calibration aligns the stringency of controls with the level of risk. This includes differentiating requirements by impact category, defining protocols for handling exceptions, applying requirements in a graduated manner, and incorporating mechanisms for periodic reassessment.

Balanced control design emphasizes enablement alongside restrictions and communicates the rationale behind controls clearly. It considers practical implementation challenges, assessing the impact on efficiency, and integrating user experience into policy design.

Standardization with flexibility includes maintaining core requirement consistency, allowing for contextual adaptation, providing guidance for local implementation, accommodating innovation, and enabling continuous evolution of policies over time.

Effective usage policies provide clear boundaries that focus on meaningful protection without imposing unnecessary constraints that impede value creation.

Boundary Condition Definition

Digital workforce governance requires an explicit definition of boundary conditions that limit autonomous operation and trigger human intervention.

Boundary type categories

1. **Authority limitations**

- Decision types requiring human approval
- Financial commitment thresholds
- Representation constraint definition
- Commitment duration restrictions
- Responsibility assumption boundaries

2. **Activity constraints**

- Prohibited action categories
- System access limitations
- Interaction restriction parameters
- Content generation boundaries
- Tool usage constraints

3. **Performance thresholds**

- Confidence level requirements
- Error rate tolerance limits
- Response time expectations
- Consistency standards
- Accuracy minimum requirements

4. **Ethical boundaries**

- Value alignment requirements
- Fairness expectation definition
- Transparency obligation parameters
- Harm avoidance constraints
- Respect and dignity standards

Boundary Type Categories

Authority limitations

Decision types requiring human approval, financial commitment thresholds, representation constraint definition, commitment duration restrictions, responsibility assumption boundaries.

Prohibited action categories, system access limitations, interaction restriction parameters, content generation boundaries, tool usage constraints.

Activity constraints

Confidence level requirements, error rate tolerance limits, response time expectations, consistency standards, accuracy minimum requirements.

Performance thresholds

Value alignment requirements, fairness expectation definition, transparency obligation parameters, harm avoidance constraints, respect and dignity standards.

Ethical boundaries

Implementation methodology elements

1. **Explicit codification**

- Clear boundary specification
- Measurable threshold definition
- Unambiguous interpretation capability
- Comprehensive coverage
- Accessible documentation

2. **Technical enforcement**

- Automated constraint implementation
- Verification mechanism development
- Override protocol definition
- Violation detection capabilities
- Logging and alerting systems

3. **Education and communication**

- User awareness development
- Designer and developer training
- Rationale transparency
- Consequence explanation
- Feedback mechanism provision

Organizations with well-defined boundary conditions typically achieve an appropriate balance between autonomous operation and human oversight, enabling digital workers to operate efficiently within clear constraints.

Exception Handling Procedures

Even the most comprehensive policies cannot anticipate all situations. Effective governance requires structured approaches for managing exceptions to established guidelines.

Exception category framework

1. **Emergency exceptions**

- Urgent situation response
- Temporary constraint relaxation
- Business continuity enablement
- Time-limited authorization
- Extraordinary circumstance management

2. **Innovation enablement exceptions**

- Controlled experimentation allowance
- Pilot implementation authorization
- Time-bounded exploration permission
- Limited scope restriction relaxation
- Supervised boundary extension

3. **Special case exceptions**

- Unique situation accommodation
- Context-specific requirement modification
- Alternative compliance approach acceptance
- Compensating control substitution
- Non-standard implementation approval

Exception Categories

Emergency Exceptions	Innovation Exceptions	Special Case Exceptions
Urgent situation response and business continuity.	Controlled experimentation and pilot implementation.	Unique situation accommodation and alternative compliance.
1	2	3

Exception process elements

1. **Request and justification**

- Standard submission format
- Business case articulation
- Risk assessment inclusion
- Mitigation plan development
- Benefit explanation requirements

2. **Review and approval**

- Appropriate authority determination
- Analysis criteria standardization
- Multi-perspective evaluation
- Precedent consideration
- Documentation requirements

3. **Implementation and monitoring**

- Scope and duration limitation
- Enhanced oversight implementation
- Regular reassessment scheduling
- Success criteria definition
- Sunset provision establishment

4. **Knowledge integration**

- Pattern recognition across exceptions
- Policy evolution consideration
- Organizational learning facilitation
- Common need identification
- Systematic enhancement implementation

Exception handling processes must enable appropriate flexibility while maintaining governance integrity. Effective processes focus on managed deviation rather than policy circumvention or rigid enforcement regardless of circumstances.

Policy Evolution and Lifecycle Management

Digital workforce policies must evolve as technologies, capabilities, risks, and organizational needs change. Effective governance includes systematic approaches to policy lifecycle management.

Evolution trigger categories

1. **Technology advancement**

- Capability expansion implications
- New risk emergence recognition
- Obsolete constraint identification
- Integration opportunity emergence

- Tool ecosystem evolution

2. **Organizational learning**

- Implementation experience integration
- Exception pattern recognition
- User feedback incorporation
- Performance insight application
- Value optimization opportunities

3. **External environment changes**

- Regulatory requirement evolution
- Competitive landscape shifts
- Customer expectation changes
- Industry standard development
- Threat landscape transformation

Navigating Evolution Triggers

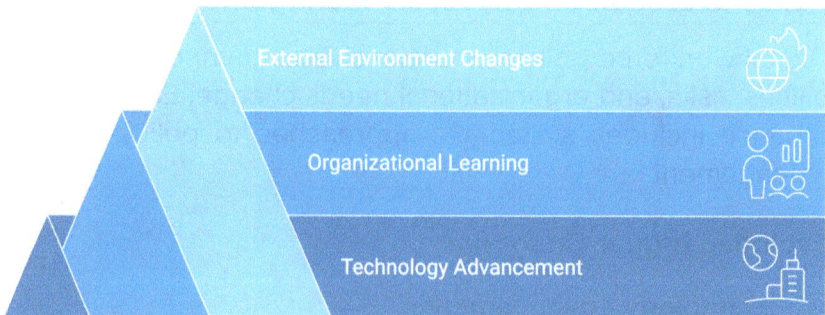

Management process components

1. **Systematic review mechanisms**

- Regular assessment cadence

- Trigger-based evaluation processes
- Stakeholder input collection
- Effectiveness measurement
- Gap and opportunity identification

2. **Controlled update approaches**

- Version management procedures
- Change impact assessment
- Communication planning
- Implementation timing considerations
- Training and awareness updates

3. **Balance maintenance**

- Protection and enablement recalibration
- Burden and benefit reassessment
- Constraint and flexibility rebalancing
- Standardization and customization adjustment
- Control and autonomy realignment

Organizations with mature policy lifecycle management capabilities typically maintain more effective digital workforce governance over time, adapting to changing circumstances while preserving core principles and protections.

Cross-functional Oversight: Establishing Effective Governance Structures

Effective digital workforce governance requires organizational structures that enable appropriate oversight, coordination, and decision-making across functional boundaries.

Committee Composition and Charter Development

Most organizations benefit from dedicated governance bodies that bring together diverse perspectives to guide digital workforce implementation and operation.

Typical committee types

1. **Executive steering committee**

- Strategic direction setting
- Resource allocation approval
- Cross-functional alignment facilitation
- Major issue resolution
- Overall accountability maintenance

2. **Technical governance committee**

- Architecture and standard definition
- Technology selection guidance
- Integration approach oversight
- Technical risk management
- Implementation quality assurance

3. **Ethics and responsible use committee**

- Value alignment verification
- Ethical implication assessment
- Fairness and bias evaluation
- Transparency standard definition
- Societal impact consideration

4. **Operational oversight committee**

- Performance monitoring
- Enhancement prioritization

- Issue management coordination
- Continuous improvement direction
- User experience optimization

Committee design elements

Membership composition focuses on assembling the right mix of individuals to ensure effective functioning. This includes ensuring appropriate functional representation, including necessary expertise, aligning membership with decision-making authority, optimizing committee size for efficiency, and incorporating diverse perspectives.

Charter development establishes the foundation for the committee's role and functioning. It involves clearly defining the purpose and scope, articulating specific responsibilities, specifying authority boundaries, clarifying relationships with other governance bodies, and outlining the operating mechanisms.

Operational process design ensures the committee runs smoothly and productively. This includes determining meeting frequency, managing agendas effectively, specifying decision-making procedures, setting documentation and communication requirements, and establishing mechanisms for evaluating committee performance.

The most effective governance committees balance representation with efficient decision-making, providing meaningful oversight without creating bureaucratic barriers to implementation progress.

Review Processes and Cadences

Systematic review processes represent a core mechanism for governance bodies to fulfill their oversight responsibilities effectively.

Review type framework

1. **Strategic portfolio reviews**

- Investment alignment assessment
- Value realization evaluation
- Strategic direction recalibration
- Resource allocation optimization
- Cross-initiative synergy identification

2. **Implementation stage-gate reviews**

- Readiness assessment for progression
- Requirement fulfillment verification
- Risk management adequacy evaluation
- Resource and timeline validation
- Go/no-go decision making

3. **Operational performance reviews**

- Key metric evaluation
- Issue pattern identification
- Enhancement opportunity prioritization
- User feedback consideration
- Comparative benchmark assessment

4. **Risk and compliance reviews**

- Control effectiveness verification
- New risk identification
- Regulatory compliance confirmation
- Incident response evaluation
- Improvement recommendation development

Review Type Framework

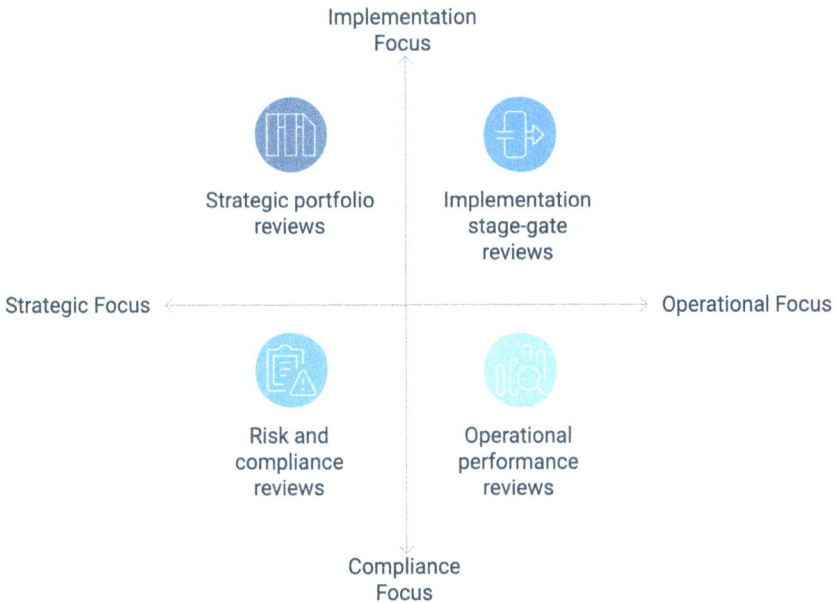

Implementation Focus

Strategic portfolio reviews

Implementation stage-gate reviews

Strategic Focus ←——————————————→ Operational Focus

Risk and compliance reviews

Operational performance reviews

Compliance Focus

Review process design considerations

Frequency calibration ensures reviews happen at the right intervals based on the nature of the content. This involves determining cadence based on impact, adjusting for process maturity, considering the volume and speed of activity, assessing sensitivity to changes, and balancing it with available resources.

Content and format optimization make the review process effective and efficient. Key steps include specifying what information is needed, standardizing how it's presented, keeping focus on the most important issues, structuring the process to support decision-making, along with continuous improvement mechanisms.

Outcome integration connects the review process to ongoing improvement. This means tracking actions that result from the review, verifying that decisions are implemented, integrating lessons learned, identifying patterns across reviews, and using findings to drive systemic improvements.

Effective review processes provide meaningful oversight that drives improvement rather than creating performative compliance exercises that consume resources without adding value.

Escalation Pathways and Resolution Mechanisms

Effective governance includes clear processes for escalating issues and conflicts to appropriate levels for resolution when normal processes prove inadequate.

Escalation trigger categories

1. **Decision deadlocks**

- Cross-functional disagreements
- Competing priority conflicts
- Resource allocation disputes
- Approach selection impasses
- Risk tolerance interpretation differences

2. **Performance issues**

- Persistent quality problems
- Significant deviation from expectations
- Recurring failure patterns
- User satisfaction deterioration
- Value delivery shortfalls

3. **Risk and compliance concerns**

- Control effectiveness questions
- Policy interpretation conflicts
- Potential violation identification

- Emerging risk recognition
- External requirement changes

4. **Strategic alignment questions**

- Direction drift identification
- Competing objective tensions
- Investment reprioritization needs
- Capability gap recognition
- External environment shifts

Escalation Trigger Categories

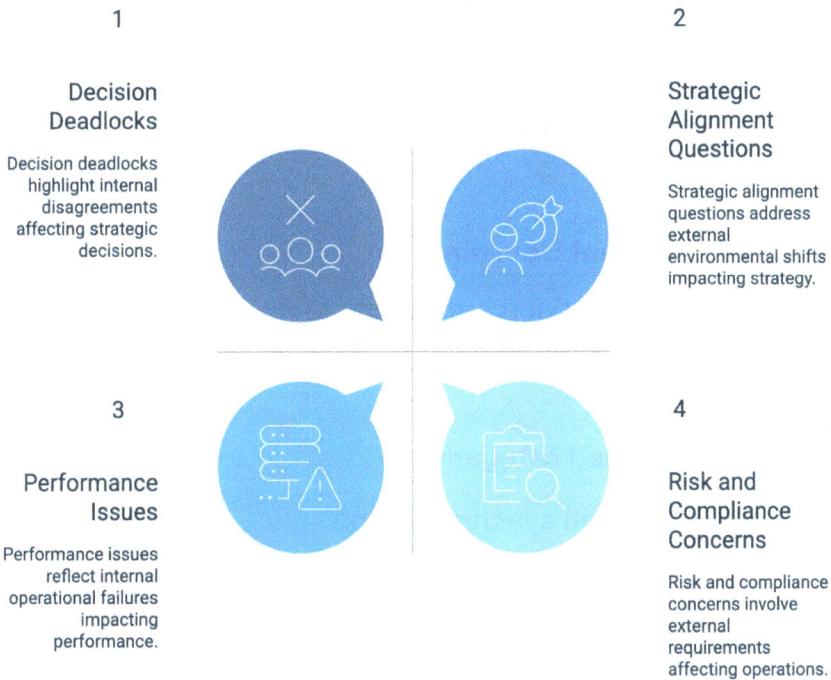

1

Decision
Deadlocks

Decision deadlocks
highlight internal
disagreements
affecting strategic
decisions.

2

Strategic
Alignment
Questions

Strategic alignment
questions address
external
environmental shifts
impacting strategy.

3

Performance
Issues

Performance issues
reflect internal
operational failures
impacting
performance.

4

Risk and
Compliance
Concerns

Risk and compliance
concerns involve
external
requirements
affecting operations.

Resolution process elements

Structured escalation paths define a clear progression for handling issues. This includes matching authority levels to issue types, enabling bypass mechanisms for urgent matters, ensuring cross-functional representation at each stage, and clearly specifying who holds final decision-making authority.

Issue documentation requirements focus on creating consistent and informative records. This involves using a standard format, providing necessary context and background, including multiple perspectives when relevant, adding recommendations, and outlining the expected impact of the issue.

Resolution approach options offer different ways to address and resolve issues. These include facilitating collaborative problem-solving, applying decision-forcing mechanisms when needed, making mediation available, incorporating external viewpoints, and using temporary solutions with ongoing monitoring when a full resolution is not immediately possible.

Organizations with well-defined escalation and resolution processes address issues more efficiently, reduce organizational friction, and maintain implementation momentum even when challenges arise.

Cross-Functional Coordination

Effective digital workforce oversight requires ongoing coordination across organizational functions to align activities and share insights.

Coordination need categories

1. **Implementation synchronization**

- Cross-initiative timing alignment
- Dependency management
- Resource conflict resolution
- Methodology consistency
- Knowledge transfer facilitation

2. **Knowledge and learning sharing**

- Best practice dissemination
- Issue pattern recognition
- Solution approach exchange
- Innovation distribution
- Failure lesson sharing

3. **Stakeholder management alignment**

- Communication consistency
- Expectation management coordination
- Change sequencing optimization
- Training approach alignment
- User experience consistency

Coordination options

1. **Communities of practice**

- Function-specific knowledge forums
- Practitioner connection facilitation
- Experience exchange structures
- Methodology evolution collaboration
- Peer support networks

2. **Integration role establishment**

- Dedicated coordination positions
- Cross-functional responsibility assignment
- Bridge-building accountability
- Information flow facilitation
- Alignment verification duty

3. **Systematic information sharing**

- Regular cross-team sessions

- Standardized update mechanisms
- Shared knowledge repositories
- Transparent project visibility
- Automated notification systems

Strategies for Unified Collaboration

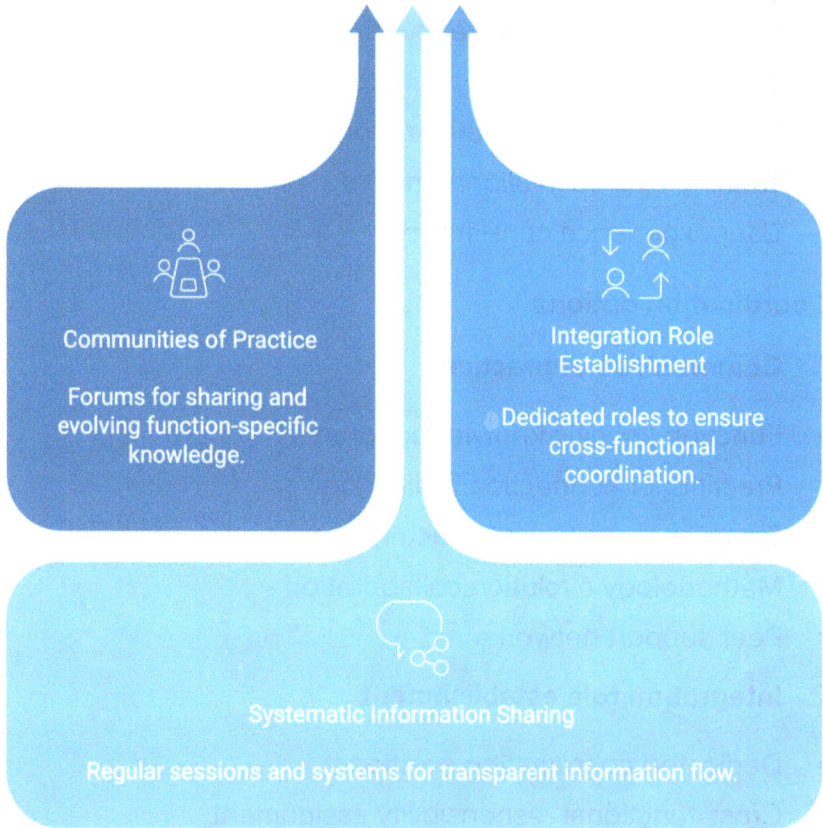

Communities of Practice

Forums for sharing and evolving function-specific knowledge.

Integration Role Establishment

Dedicated roles to ensure cross-functional coordination.

Systematic Information Sharing

Regular sessions and systems for transparent information flow.

The most effective coordination approaches balance formal mechanisms with cultural enablement, creating both structures and norms that support ongoing alignment across organizational boundaries.

External Stakeholder Management: Addressing Broader Ecosystem Concerns

Digital workforce governance extends beyond internal organizational boundaries to include relationships with customers, partners, regulators, and society at large. Effective governance addresses these external dimensions explicitly.

Customer Communication Strategies

Organizations must thoughtfully manage how they communicate with customers about digital worker capabilities, limitations, and usage to build trust and set appropriate expectations.

Communication dimension framework

1. **Transparency level determination**

- Disclosure scope decisions
- Interaction identification requirements
- Capability explanation approach
- Limitation communication strategy
- Human involvement clarification

2. **Experience design considerations**

- Transition experience optimization
- Choice and control provision
- Feedback mechanism implementation
- Preference management capabilities
- Personalizing experience within boundaries

3. **Value communication approaches**

- Benefit explanation strategies
- Improvement demonstration

- Comparative advantage illustration
- Future capability roadmap sharing
- User story and example utilization

Communication Dimension Framework

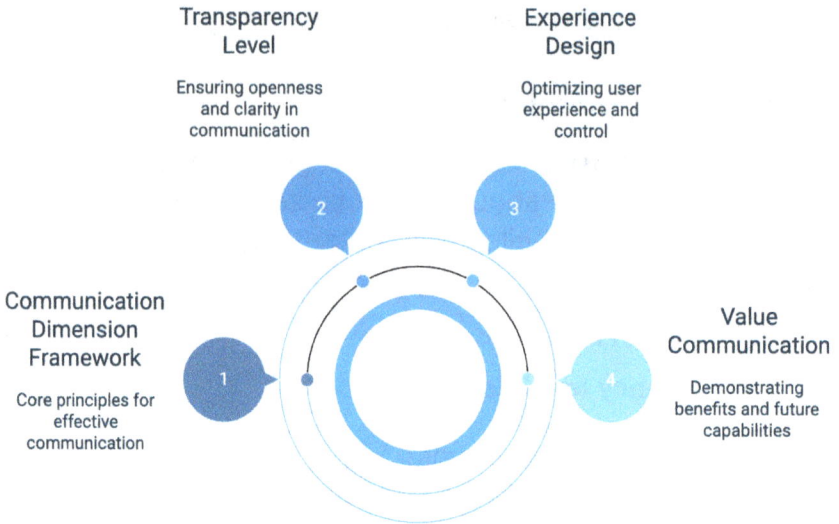

Transparency Level
Ensuring openness and clarity in communication

Experience Design
Optimizing user experience and control

Communication Dimension Framework
Core principles for effective communication

Value Communication
Demonstrating benefits and future capabilities

Implementation approach elements

Segmented communication strategy tailors messaging based on audience characteristics. This involves using different approaches for each customer type, adapting to user preferences, calibrating the level of detail, optimizing timing for each segment, and selecting communication channels suited to the recipient.

Progressive disclosure methods present information in manageable layers. This includes using a layered information structure, allowing access to detail based on user interest, providing explanations just in time, presenting context-relevant content, and enabling users to explore further as needed.

Experience consistency assurance ensures a unified and coherent user experience. This includes verifying brand alignment, coordinating messages across channels, standardizing visual and verbal cues, maintaining consistent interaction patterns, and aligning user expectations with actual delivery.

Organizations with sophisticated customer communication strategies typically achieve higher trust, satisfaction, and adoption than those that either over-promise capabilities or inadequately prepare customers for digital worker interactions.

Supplier and Partner Relationship Management

Digital workforce implementations often involve external vendors, technology providers, and implementation partners, requiring governance approaches that extend across organizational boundaries.

Relationship type categories

1. **Technology provider relationships**

- Foundation model suppliers
- Platform and infrastructure providers
- Tool and component vendors
- Integration solution partners
- Specialized capability licensors

2. **Implementation partner relationships**

- Strategy consulting providers
- Design and development partners
- Integration specialists
- Change management consultants
- Training and enablement providers

3. **Operational support relationships**

- Ongoing maintenance providers
- Monitoring and management partners

- Enhancement and optimization specialists
- Support service suppliers
- Content and knowledge providers

Governance approach elements

1. **Selection and evaluation frameworks**

- Capability assessment methodologies
- Risk evaluation approaches
- Cultural alignment consideration
- Value alignment verification
- Long-term viability assessment

2. **Contractual governance mechanisms**

- Performance specification approaches
- Liability and responsibility allocation
- Intellectual property rights management
- Data usage and ownership definition
- Exit and transition provisions

3. **Operational governance processes**

- Regular review cadence establishment
- Performance measurement frameworks
- Issue management procedures
- Change control processes
- Continuous improvement mechanisms

Governance Approach Elements

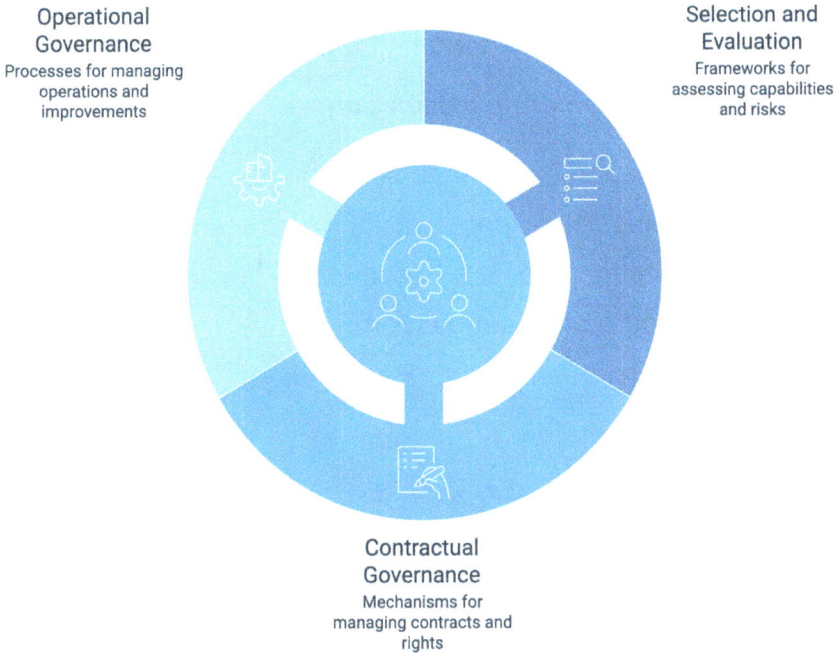

Operational Governance
Processes for managing operations and improvements

Selection and Evaluation
Frameworks for assessing capabilities and risks

Contractual Governance
Mechanisms for managing contracts and rights

The most effective external relationship governance balances clear accountability with collaborative partnership. This creates relationships that deliver value while appropriately managing risk across organizational boundaries.

Regulatory Engagement Approaches

As regulatory attention to AI and digital workers increases, organizations must develop systematic approaches to engage with regulatory bodies and contribute to the evolving governance landscape.

Engagement purpose categories

1. **Compliance assurance**

- Requirement interpretation clarification

- Implementation approach validation
- Examination preparation and support
- Issue remediation alignment
- Ongoing compliance demonstration

2. **Policy development participation**

- Perspective and experience sharing
- Technical feasibility input
- Implementation impact assessment
- Practical alternative suggestion
- Balance consideration advocacy

3. **Relationship building**

- Understanding development
- Trust establishment
- Communication channel creation
- Collaborative approach building
- Shared objective identification

Strategy implementation elements

1. **Monitoring and intelligence**

- Regulatory development tracking
- Enforcement trend analysis
- Peer and industry practice assessment
- Global perspective maintenance
- Interpretation insight development

2. **Engagement approach optimization**

- Appropriate representative selection
- Message development and consistency

- Timing and forum choice
- Coalition building where beneficial
- Balanced perspective presentation

3. **Organizational alignment**

- Cross-functional coordination
- Consistent position development
- Implementation impact consideration
- Commercial and compliance balance
- Risk and opportunity evaluation

Strategy Implementation Elements

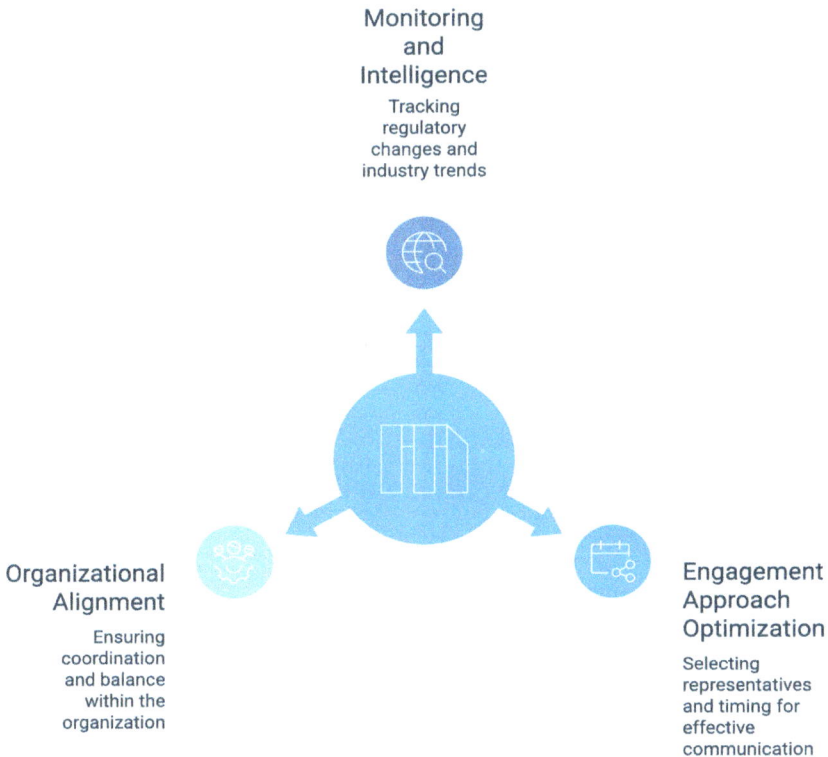

Monitoring and Intelligence

Tracking regulatory changes and industry trends

Organizational Alignment

Ensuring coordination and balance within the organization

Engagement Approach Optimization

Selecting representatives and timing for effective communication

Organizations that develop sophisticated regulatory engagement capabilities typically navigate the evolving landscape more effectively. They are able to influence development toward practicable approaches while maintaining appropriate compliance.

Public and Community Responsibility

Digital workers can have broader societal impacts beyond immediate stakeholders, requiring governance approaches that consider larger community responsibility dimensions.

Responsibility dimension framework

1. **Economic impact consideration**

- Job and role evolution effects
- Skill requirement shifts
- Opportunity access questions
- Value distribution concerns
- Transition management approaches

2. **Environmental sustainability**

- Energy consumption implications
- Resource utilization efficiency
- Carbon footprint management
- E-waste and lifecycle considerations
- Sustainable design principles

3. **Societal well-being factors**

- Inclusivity and accessibility
- Digital divide implications
- Vulnerable population considerations
- Cultural sensitivity aspects

- Community relationship impacts

Implementation approach elements

Impact assessment methodologies are essential for evaluating how a system or solution will affect different groups and contexts. This includes using frameworks that consider multiple dimensions of impact, consulting with stakeholders, thinking through long-term implications, identifying unintended consequences, and regularly reassessing as the system evolves.

Design and operation principles focus on ensuring the solution is inclusive, fair, and beneficial. This means applying inclusive design practices, meeting accessibility standards, verifying fairness throughout the process, optimizing for community benefit, and prioritizing harm minimization wherever possible.

Engagement and transparency are critical for trust and adaptability. Effective approaches include mechanisms for community consultation, clear communication about potential impacts, integrating feedback into improvements, adapting based on expressed concerns, and maintaining an open, ongoing dialogue with stakeholders.

Organizations with mature approaches to societal responsibility typically create more sustainable digital workforce implementations that foster stakeholder trust while minimizing negative impacts.

Conclusion: Governance as Enabler Rather Than Constraint

Effective digital workforce governance is a crucial success factor for organizations seeking to capture the full potential of agentic AI while managing associated risks. The most successful approaches recognize governance as an enabler of sustainable value creation rather than merely a constraint on innovation.

Several key principles distinguish successful governance:

1. **Clear accountability with appropriate autonomy**: Establishing unambiguous responsibility while providing adequate freedom to innovate within defined boundaries.

2. **Balanced controls calibrated to risk**: Implementing controls proportionate to actual risk rather than applying uniform constraints regardless of context.

3. **Cross-functional integration with decision efficiency**: Bringing diverse perspectives together while maintaining practical decision velocity.

4. **Adaptability within principled frameworks**: Enabling evolution to address changing needs while preserving core governance principles.

5. **External stakeholder inclusion with practical focus**: Considering broader ecosystem perspectives without becoming paralyzed by excessive consultation.

Principles of Effective Governance

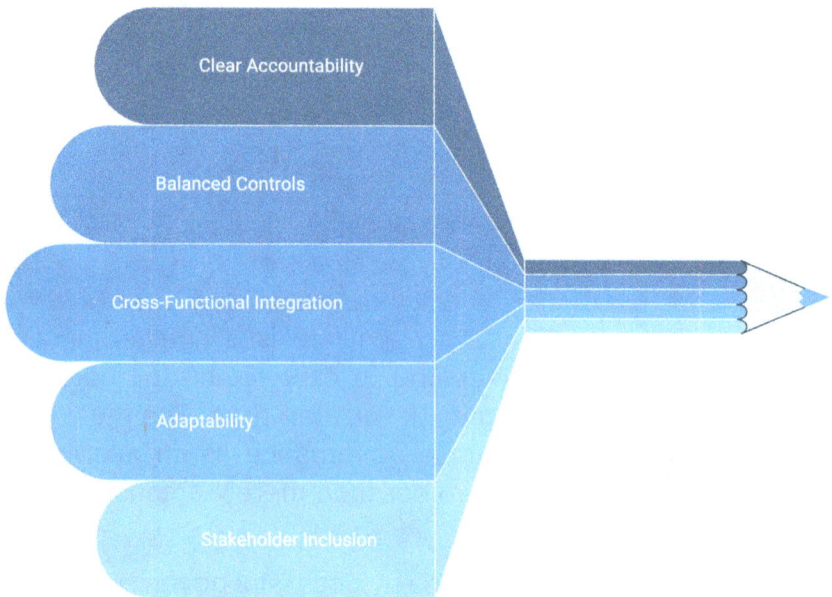

Clear Accountability

Balanced Controls

Cross-Functional Integration

Adaptability

Stakeholder Inclusion

As digital workforces continue to expand in organizational impact, the sophistication of governance approaches will become essential. It will increasingly differentiate organizations that capture sustainable value from those that either miss opportunities through excessive constraint or create unsustainable risks through inadequate oversight.

PART
FOUR
THE CURRENT LANDSCAPE AND FUTURE DIRECTIONS

CHAPTER 10:
SURVEY FINDINGS - THE STATE OF AGENTIC AI ADOPTION

<table>
<tr><td align="center">TL;DR:</td></tr>
</table>

- 89.1% of organizations are already exploring or implementing agentic AI, with 49.1% having active production deployments—far beyond commonly assumed early experimental phases.

- 71.1% view agentic AI as essential for competitive positioning rather than merely operational efficiency, with 79.8% expecting industry transformation within five years.

- IT and security (53.9%) and customer service (46.8%) lead functional adoption, following a risk-calibrated approach from lower-risk to more complex applications.

- Technical integration complexity (48.5%), skills/talent gaps (42.1%), and data quality issues (40.6%) are the primary barriers, while unclear ROI ranks lowest (9.9%), indicating value recognition despite implementation difficulties.

- Organizations face a limited window to establish competitive advantages through superior agentic AI implementation before basic capabilities become commoditized across industries.

Understanding Agentic AI Adoption

As organizations navigate the digital workforce transformation, understanding the current landscape of agentic AI adoption provides crucial context for strategic decision-making.

Findings from our comprehensive survey conducted in June 2025 offer insights into implementation patterns, challenges, success factors, and strategic approaches to guide leaders in their digital transformation journeys.

Research Methodology: Establishing Credibility and Context

To provide a comprehensive view of the current state of agentic AI adoption, we conducted a detailed survey designed to capture both the breadth and depth of implementation experiences across diverse organizational contexts.

Survey Design and Execution

Our research employed a structured methodology combining quantitative measurement with nuanced categorical insights to understand organizational experiences with digital workers.

Survey scope and reach

The survey included 442 business professionals from a broad spectrum of roles, industries, and organizational sizes:

- **Role distribution**: 26.9% executive management (President, CEO, COO, CFO), 33.9% middle management and functional leadership, 39.2% individual contributors and specialists

- **Company size range**: From small organizations (1-10 employees: 9.3%) to large enterprises (1000+ employees: 39.4%), with strong representation across the spectrum

- **Revenue distribution**: Organizations ranging from under $5 million (18.9%) to over $1 billion in annual revenue, providing insights across different resource levels

- **Industry representation**: Broad coverage including High tech (15.9%), financial services (11.7%), business/professional services (10.4%), construction (10.4%), and other major sectors

Participant characteristics

The respondent profile reveals a mature, experienced population well-positioned to assess organizational AI adoption:

1. **Age distribution**: 70.7% between ages 31-50, representing experienced professionals in decision-making roles
2. **Employment status**: 89.6% employed full-time, with 6.4% self-employed, ensuring insights from operational rather than theoretical perspectives
3. **Functional diversity**: Strong representation from executive management, IT/technology, operations, sales and marketing, and other key functions

This diverse participant population enables insights that span different organizational contexts while identifying common patterns that transcend specific circumstances.

Data Collection Approach

Our methodology captured actionable insights rather than merely descriptive statistics:

Adoption maturity assessment: Multi-stage framework measuring progression from initial exploration through enterprise-wide implementation

Implementation experience mapping: Detailed examination of functional areas, challenges, success factors, and organizational impacts

Strategic perspective analysis: Understanding how organizations view agentic AI within their broader competitive and transformation strategies

Forward-looking insights: Expectations about industry transformation and future implementation priorities

This comprehensive approach enables presentation of findings that combine statistical rigor with practical understanding of the complex realities organizations face in their digital workforce journeys.

The Maturity Surprise: How Far Along Adoption Actually Is

The most significant finding from our research challenges common assumptions about agentic AI adoption being in early experimental stages. The data reveals a market that has moved substantially beyond initial exploration into active implementation.

Adoption Stages: Beyond Early Exploration

The distribution of organizations across adoption stages presents a striking picture of market maturity:

- **Not yet exploring agentic AI**: 10.9% (48 organizations)
- **Researching potential applications**: 16.3% (72 organizations)
- **Planning initial implementation**: 15.2% (67 organizations)
- **Pilot projects underway**: 8.6% (38 organizations)
- **Multiple implementations in production**: 28.7% (127 organizations)
- **Enterprise-wide implementation**: 20.4% (90 organizations)

The headline insight: Nearly 9 out of 10 organizations (89.1%) have begun exploring agentic AI, with only 10.9% not yet engaged with these technologies.

More significantly, 57.7% of organizations have moved beyond research and planning phases into active implementation—pilot projects, production deployments, or enterprise-wide rollouts.

This suggests that agentic AI adoption has reached a tipping point where implementation experience, rather than theoretical exploration, dominates organizational thinking.

Active Implementation Reaches Critical Mass

When we examine organizations with actual operational deployments (combining "multiple implementations in production" and "enterprise-wide implementation"), we find that 49.1% of surveyed organizations, *nearly half*, have moved beyond experimentation into business-critical implementations.

This finding contradicts common narratives suggesting agentic AI remains primarily experimental. Instead, it reveals a market where:

- **Only a small number of Companies in the Proof-of-concept phase**: Only 8.6% remain in pilot phases
- **Production deployment is mainstream**: 28.7% have multiple implementations running
- **Enterprise transformation is underway**: 20.4% have achieved enterprise-wide implementation

Implications for Market Dynamics

This maturity level creates several important implications:

Competitive pressure intensifies: With nearly half of organizations having active implementations, competitive advantage increasingly depends on execution quality rather than first-mover status.

Learning curve acceleration: The substantial population of organizations with implementation experience creates opportunities for best practice sharing and accelerated learning across industries.

Vendor ecosystem maturation: High adoption rates drive demand for more sophisticated solutions, professional services, and specialized talent, accelerating ecosystem development.

Implementation risk shifts: As adoption moves from experimental to operational, focus shifts from proving concept viability to managing integration complexity, navigating change management, scaling challenges, and overseeing ongoing agent management.

The data suggests we have entered a new phase where agentic AI implementation moves from "whether" to "how" for most organizations. Strategic advantage is now flowing to those who execute most effectively rather than those who simply adopt first.

Implementation Patterns: Where and How Organizations Are Deploying

Understanding where organizations are deploying agentic AI reveals clear patterns in implementation strategy, with certain functional areas emerging as early leaders while others remain largely untapped.

Functional Area Leadership: IT and Customer-Facing Functions Lead

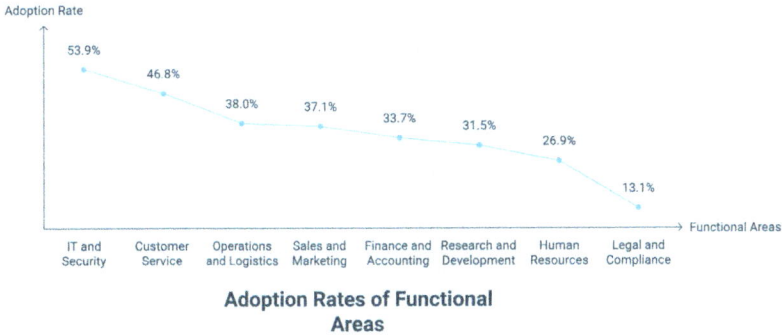

Adoption Rates of Functional Areas

The survey data reveals distinct tiers in functional area adoption:

Tier 1 - Leading implementation areas (40%+ adoption):

- **IT and security**: 53.9% (238 organizations) - The clear leader
- **Customer service**: 46.8% (207 organizations) - Strong second position

Tier 2 - Mainstream implementation areas (30-40% adoption):

- **Operations and logistics**: 38.0% (168 organizations)
- **Sales and marketing**: 37.1% (164 organizations)
- **Finance and accounting**: 33.7% (149 organizations)
- **Research and development (R&D)**: 31.5% (139 organizations)

Tier 3 - Emerging implementation areas (Below 30% adoption):

- **Human resources**: 26.9% (119 organizations)
- **Legal and compliance**: 13.1% (58 organizations)

Understanding Implementation Preferences

Several factors explain these adoption patterns:

IT and security leadership (53.9%)

The dominance of IT implementation reflects several advantages.

There's a high level of technical comfort, as IT teams already possess the skills and familiarity required to implement AI systems. IT's ownership of infrastructure ensures direct control over the foundational systems needed for successful AI implementation.

AI initiatives in IT also benefit from being tested in a controlled environment, where there's less direct exposure to customers, reducing risk during initial deployment and refinement. The presence of clear use cases, such as monitoring, automation, threat detection, and system optimization, makes implementation more straightforward.

Customer service success (46.8%)

High adoption in customer service demonstrates organizational confidence in AI's ability to handle complex interactions.

The scalability pressure in this function, driven by constant demand for responsiveness and availability, makes it especially suitable for AI augmentation. Adoption is further supported by measurable outcomes, with clear KPIs such as response time, resolution rate, and customer satisfaction offering tangible evidence of impact.

There's also a defined evolution path, as many organizations move from basic chatbots to more advanced, agentic interactions. Finally, the cost impact is a strong driver: AI enables both significant operational cost savings and improvements in overall service quality.

Mid-tier functions (30-40%)

Operations, Sales, Marketing, Finance, and R&D show substantial adoption, reflecting several enabling factors.

These areas offer process automation opportunities through structured workflows that lend themselves well to intelligent automation. They also operate in data-rich environments, giving AI systems the input they need to perform effectively.

These functions face performance pressure—driven by competition—to enhance both efficiency and effectiveness, imploring them to innovate. Finally, there is ROI clarity, as the value proposition for AI in these domains is typically well understood and easy to articulate.

Notable Adoption Gaps

The low adoption in certain areas reveals important constraints and opportunities:

Legal and compliance lag (13.1%)

The lowest adoption rate is largely due to risk aversion, as legal and compliance teams operate in high-stakes environments with strict regulatory demands that encourage conservative approaches.

Complexity requirements also present a barrier. Legal reasoning and nuanced compliance interpretation often exceed current AI capabilities. There are also liability concerns, particularly around who holds responsibility for AI-generated advice or decisions.

Regulatory uncertainty regarding how AI should be governed in these sensitive areas adds to the hesitation.

Human resources moderate adoption (26.9%)

HR's position suggests cautious but growing implementation.

Sensitive applications like hiring, performance reviews, and employee relations require thoughtful, transparent AI deployment. There's also heightened focus on fairness concerns, as bias and discrimination risks are especially significant in HR use cases.

At the same time, there's the value of personal touch, recognizing that human interaction remains critical in many people-related functions.

Compliance complexity, due to varying employment laws and equal opportunity mandates, adds another layer of challenge to widespread AI implementation in HR.

Strategic Implementation Insights

These patterns reveal several strategic insights:

Function-first strategy: Organizations are implementing agentic AI function by function rather than enterprise-wide, allowing for controlled experimentation and learning accumulation.

Risk-calibrated approach: Higher adoption in lower-risk areas (IT, operations) compared to high-stakes functions (legal, some HR applications) demonstrates thoughtful risk management.

Value-driven selection: Strong adoption in areas with clear value propositions (customer service efficiency, IT automation) suggests organizations prioritize demonstrable ROI.

Evolution pathway: The pattern suggests a natural progression from technical functions toward more complex business and judgment-intensive applications as capabilities and confidence develop.

AI Adoption Strategies

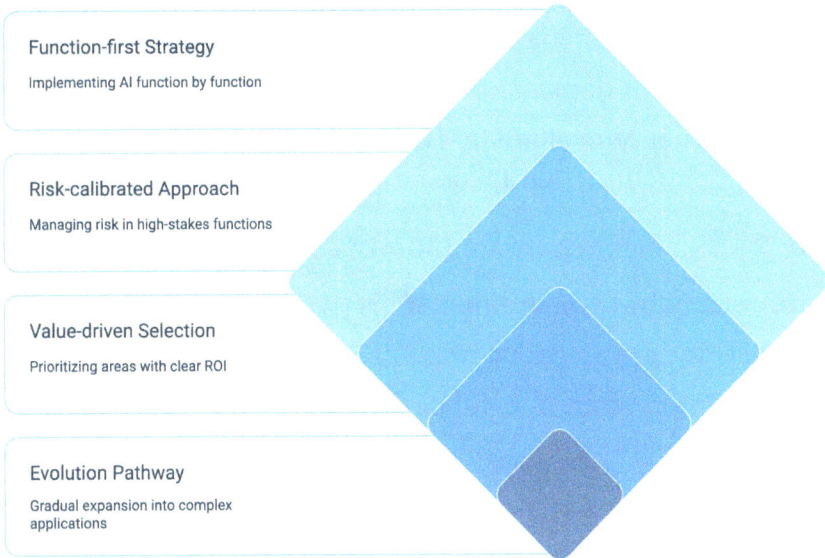

Function-first Strategy

Implementing AI function by function

Risk-calibrated Approach

Managing risk in high-stakes functions

Value-driven Selection

Prioritizing areas with clear ROI

Evolution Pathway

Gradual expansion into complex applications

Those considering agentic AI implementation can use these patterns to benchmark their approaches, identify likely starting points, and understand the typical progression of cross-functional adoption within similar organizations.

The Execution Challenge: Barriers Preventing Faster Adoption

Adoption rates demonstrate substantial interest in agentic AI, but the challenges organizations face reveal important barriers that separate successful implementation from good intentions. Understanding these obstacles provides crucial insights for implementation planning and risk management.

Challenge Hierarchy: Technical Integration Leads

The survey identified a clear hierarchy of implementation challenges:

Tier 1 - Primary barriers (40%+ of organizations):

- **Technical integration complexity**: 48.5% (196 organizations) - The dominant challenge
- **Skills/talent gaps**: 42.1% (170 organizations) - Close second
- **Data quality/availability issues**: 40.6% (164 organizations) - Infrastructure foundation

Tier 2 - Significant barriers (30-40% of organizations):

- **Cost/budget constraints**: 36.1% (146 organizations)
- **Security/privacy concerns**: 35.6% (144 organizations)
- **Regulatory/compliance concerns**: 30.9% (125 organizations)

Tier 3 - Secondary barriers (Below 30% of organizations):

- **Change management/employee resistance**: 24.8% (100 organizations)
- **Unclear ROI/business Case**: 9.9% (40 organizations) - Notably low

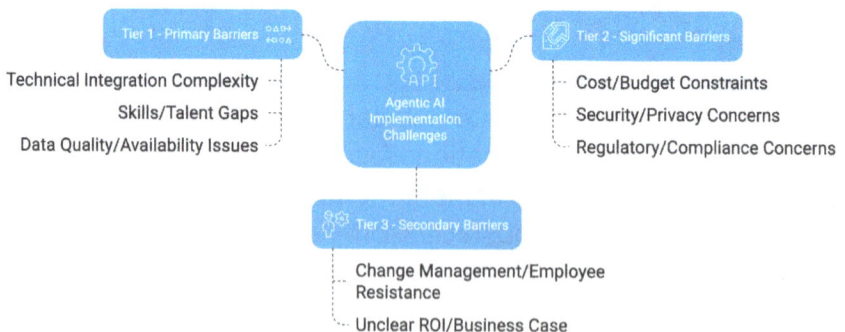

Agentic AI Implementation Challenges

Technical Integration: The Dominant Challenge

The fact that technical integration complexity ranks as the top challenge (48.5%) reveals several important insights:

Legacy system reality: Most organizations must integrate agentic AI with existing enterprise systems, databases, and workflows that were not designed for AI integration.

Architecture misalignment: Traditional enterprise architectures often lack the flexibility, data accessibility, and real-time processing capabilities that agentic AI systems require for optimal performance.

Integration expertise gap: Organizations often possess either AI expertise or enterprise integration expertise, but rarely both in sufficient depth to manage complex integration projects effectively.

Underestimated complexity: Many organizations approach agentic AI implementation, focusing on the AI capabilities while underestimating the system integration requirements necessary for production deployment.

Skills and Talent: The Human Capital Constraint

The second-highest challenge (skills/talent gaps at 42.1%) highlights the critical role of human capital in successful implementation:

Multidisciplinary requirements: Successful agentic AI implementation requires a combination of AI expertise, domain knowledge, integration skills, change management capabilities, and business analysis—a rare combination in single individuals or teams.

Market competition: High demand for AI talent creates competitive recruitment environments, particularly challenging for organizations outside technology centers.

Internal development time: Internal capability building through training and development requires substantial time investment that may not align with implementation timelines.

Evolving skill requirements: The rapid pace of AI development means required skills continue to evolve, making both recruitment and development moving targets.

Data Infrastructure: The Foundation Challenge

Data quality and availability issues (40.6%) reflect the requirement for high-quality information to enable effective agentic AI systems:

Data preparation requirements: Agentic AI systems require clean, structured, and accessible data that many organizations have not previously needed to maintain at the required quality levels.

System fragmentation: Data often exists across multiple systems with different formats, quality standards, and access mechanisms, creating integration challenges.

Privacy and governance: The implementation of appropriate data access controls while enabling AI functionality requires advanced data governance frameworks.

Real-time requirements: Many agentic AI applications require real-time or near-real-time data access that existing systems may not support.

Cost and Security: Practical Constraints

Cost/budget constraints (36.1%) and security/privacy concerns (35.6%) show practical implementation barriers:

Investment requirements: While organizations see value in agentic AI, the substantial upfront investment required for comprehensive implementation creates budget pressure, particularly for smaller organizations.

Security architecture: The implementation of agentic AI while maintaining security standards requires new approaches to system design, access control, and monitoring that add complexity and cost.

Privacy compliance: Compliance with privacy regulations in agentic AI implementations while maintaining functionality requires careful design and ongoing monitoring.

The ROI Paradox: Clarity Despite Challenges

Unclear ROI/business case ranks lowest among challenges (9.9%), creating an important paradox: organizations see clear value in agentic AI but struggle with *execution* rather than *justification*.

This finding suggests:

Value proposition clarity: Organizations understand the potential benefits of agentic AI implementation, indicating successful thought leadership and case study dissemination within business communities.

Execution gap: The primary barriers are operational (integration, skills, data) rather than strategic (value proposition, business case), suggesting a need for implementation support.

Implementation maturity: Organizations have moved beyond questioning whether to implement agentic AI to focusing on how to implement it successfully.

Strategic Implications for Implementation

These challenge patterns provide several strategic insights:

Capability building priority: Organizations should invest in technical integration capabilities, skills development, and data infrastructure as prerequisites for successful agentic AI implementation rather than as parallel activities.

Partnership strategy: The complexity of challenges suggests that strategic partnerships with system integrators, AI specialists, and consulting firms may be more valuable than purely technology vendor relationships.

Phased implementation: The challenge hierarchy suggests staged implementation approaches that address integration complexity gradually rather than attempting comprehensive deployment.

Organizational readiness: Success depends as much on organizational infrastructure and capabilities as on AI technology selection, requiring a holistic readiness assessment and preparation.

Understanding these challenges enables organizations to develop more realistic implementation plans, appropriate resource allocation strategies, and partnership approaches that address actual barriers rather than perceived obstacles.

Strategic Positioning: How Organizations View Competitive Implications

The survey reveals sophisticated strategic thinking about agentic AI's role in competitive positioning. Organizations view these technologies as essential for future competitive advantage rather than merely operational improvements.

Strategic Perspective: Beyond Cost Reduction

Organizations demonstrate nuanced strategic thinking about agentic AI's competitive implications:

- **Competitive advantage**: 26.8% (107 organizations) - Largest single category

- **Transformative opportunity**: 24.8% (99 organizations) - Close second

- **Competitive necessity**: 19.5% (78 organizations) - Substantial group

- **Cost-cutting tool**: 18.5% (74 organizations) - Traditional view

- **Potential risk to be managed**: 6.8% (27 organizations) - Small cautious group

- **No clear strategic view yet**: 3.8% (15 organizations) - Minimal uncertainty

Advanced Strategic Thinking Dominates

Combining the top three strategic perspectives— competitive advantage, transformative opportunity, and competitive necessity—reveals that **71.1% of organizations view agentic AI as essential for competitive positioning** rather than purely operational efficiency. This suggests several important market dynamics:

Beyond operational efficiency: Only 18.5% view agentic AI primarily as a cost-cutting tool, indicating organizations recognize broader strategic potential beyond process automation.

Competitive imperative recognition: The combination of competitive advantage (26.8%) and competitive necessity (19.5%) suggests 46.3% of organizations view agentic AI as directly connected to competitive positioning.

Transformation expectations: Nearly a quarter (24.8%) view agentic AI as creating transformative opportunities, suggesting expectations for changes in business model or market structure.

Risk awareness without paralysis: Only 6.8% view agentic AI primarily as a risk to be managed, indicating organizations recognize challenges while maintaining overall optimism about potential benefits.

Competitive Advantage Positioning: The Leading Perspective

The largest single group (26.8%) views agentic AI as a competitive advantage, suggesting organizations believe they can create differentiated value through superior implementation:

Differentiation potential: Creating competitive advantages through better integration, superior user experiences, or unique applications rather than simply adopting generic AI solutions.

First-mover opportunities: Leveraging timing and execution quality to create sustainable competitive advantages, despite widespread adoption.

Capability-based competition: Viewing agentic AI as a competitive advantage implies organizations plan to compete on implementation excellence and organizational adaptation rather than technology access alone.

Value creation focus: Creating new value for customers and stakeholders rather than simply reducing costs or improving internal efficiency.

Transformative Opportunity: Change Expectation

The second-largest group (24.8%) views agentic AI as a transformative opportunity, indicating expectation of profound business model and market evolution:

Business model innovation: Enabling new ways of creating and delivering value that were previously impossible or uneconomical.

Market structure evolution: Reshaping industry competitive dynamics, customer expectations, and value chains.

Strategic reinvention: Redefining their competitive positioning rather than simply enhancing current approaches.

Long-term vision: Embracing transformative opportunity thinking implies strategic planning horizons that extend beyond immediate operational improvements to capability evolution.

Competitive Necessity: Market Forces Recognition

Nearly one in five organizations (19.5%) views agentic AI as a competitive necessity, suggesting market pressure and competitive dynamics drive implementation decisions:

Defensive positioning: Believing they must implement agentic AI to maintain competitive parity rather than create advantage.

Market evolution response: Observing competitors gaining advantages through agentic AI implementation creates pressure to follow.

Risk mitigation focus: Avoiding competitive disadvantage rather than creating superior positioning.

Industry standards evolution: Observing agentic AI capabilities becoming expected rather than differentiating in their markets.

Strategic Confidence Indicators

Several survey findings indicate high strategic confidence in agentic AI potential:

Low strategic uncertainty: Only 3.8% report no clear strategic view, suggesting most organizations have developed definitive perspectives on agentic AI's strategic role.

Minimal risk-focused thinking: Only 6.8% view agentic AI primarily as a risk to be managed, indicating organizations have moved beyond cautious evaluation to implementation focus.

Investment alignment: Strategic perspective distribution aligns with implementation patterns, suggesting organizations are investing consistently with their strategic assessments.

Implications for Competitive Dynamics

These strategic perspectives suggest several important competitive implications:

Competitive intensity: With 71.1% viewing agentic AI as competitively essential, organizations face increasing pressure to implement effectively or risk competitive disadvantage.

Differentiation challenges: As adoption becomes widespread, competitive advantage will increasingly depend on implementation excellence rather than adoption timing.

Strategic coherence requirements: Organizations must align their agentic AI investments with broader strategic positioning to capture identified competitive benefits.

Market evolution acceleration: High strategic confidence and broad adoption will likely accelerate the pace of market evolution and competitive dynamics in most industries.

The strategic sophistication demonstrated in these survey responses suggests agentic AI adoption has moved well beyond experimental phases into strategic planning processes across a broad range of organizations and industries.

Industry and Size Dynamics: Variations Across Different Contexts

While overall adoption patterns reveal significant market maturity, important variations exist across industry sectors and organizational sizes. These provide insights into contextual factors influencing the success of agentic AI implementation.

Industry Variation Patterns: Technology Leads, Traditional Sectors Follow

The survey data reveals clear industry-based adoption patterns that reflect both technological sophistication and regulatory environments:

Technology-forward industries

- **High tech**: 15.9% of survey participants, representing the largest single industry group with characteristically high adoption rates and sophisticated implementation approaches

- **Financial services/real estate/insurance**: 11.7% representation, showing strong adoption driven by data availability, process standardization, and competitive pressure

- **Telecommunications**: 5.9% representation, but typically showing advanced implementation due to technical infrastructure and customer service automation needs

Professional services industries

- **Business/professional services**: 10.4% representation, showing moderate to high adoption as firms seek to augment knowledge work and improve client service delivery
- **Construction**: 10.4% representation, with adoption focused on project management, scheduling, and resource optimization applications

Traditional manufacturing and industrial industries

- **Industrial/manufacturing**: 6.4% representation, with implementations concentrated in process optimization, quality control, and predictive maintenance
- **Wholesale/retail trade**: 6.6% representation, focusing on inventory management, customer service, and supply chain optimization

Regulated industries

- **Healthcare**: 4.0% representation, showing cautious but growing adoption due to regulatory requirements and patient safety considerations
- **Pharmaceuticals/biotechnology/life sciences**: 1.1% representation, with specialized applications in research and regulatory compliance
- **Public sector/nonprofit**: 2.0% representation, with adoption driven by efficiency requirements and citizen service improvement

Industry-Specific Implementation Characteristics

Different industries demonstrate distinct implementation approaches reflecting their unique operational requirements and constraints:

High-tech industry leadership

Technology companies typically demonstrate early adoption of agentic AI. These companies often lead with multiple enterprise-wide implementations and employ sophisticated integration strategies. They also show a greater appetite for experimentation, embracing cutting-edge capabilities ahead of other sectors.

Adoption spans both internal efficiency efforts and product enhancements, with strong in-house technical talent that allows these organizations to innovate with less reliance on external vendors or consultants.

Financial services strategic implementation

Financial institutions show a focused and risk-aware approach to agentic AI. Adoption is strong in areas like customer service, fraud detection, and compliance and implementations prioritize security and privacy, aligning with industry standards and consumer trust expectations.

These organizations also aim for differentiation through superior customer experience. These efforts are backed by substantial investments in data quality and integration infrastructure, and are always guided by rigorous compliance and risk management frameworks.

Professional services efficiency focus

Professional service firms concentrate adoption in knowledge management, research assistance, and client service, where AI augments workflows without replacing professional expertise.

The goal is to improve service delivery quality and client satisfaction while preserving trust and relationships. They prioritize professional integrity as they gradually expand from support functions to client-facing applications.

Organizational Size Impact: Resources and Complexity

Survey data reveals important relationships between organizational size and agentic AI adoption approaches:

Large enterprises (1000+ employees: 33.4% of sample)

- Higher likelihood of enterprise-wide implementations
- Greater resources for comprehensive integration projects
- More complex technical environments that require sophisticated solutions
- Established IT departments with AI implementation capabilities
- Ability to justify larger upfront investments through scale economics

Mid-size organizations (100-999 employees: 46.4% of sample)

- Focus on targeted, high-impact implementations
- Greater reliance on external partners and consultants
- Emphasis on quick wins and clear ROI demonstration
- Balance between resource constraints and competitive pressure
- Selective implementation in functions with the clearest business cases

Small organizations (1-99 employees: 20.1% of sample)

- Adoption focused on readily available solutions requiring minimal customization

- Strong preference for cloud-based, low-maintenance implementations
- Emphasis on immediate productivity improvements
- Limited technical resources requiring simple integration approaches
- Focus on competitive parity rather than differentiation through AI

Resource and Capability Implications

Organizational size significantly impacts implementation approaches:

Technical capability access

- Larger organizations are more likely to develop internal AI capabilities
- Smaller organizations rely heavily on external solutions and support
- Mid-size organizations often use hybrid approaches combining internal and external resources

Implementation complexity management

- Large organizations can manage complex, multi-phase implementations
- Small organizations require simpler, faster implementation approaches
- Mid-size organizations balance complexity and resource constraints through phased approaches

Investment patterns

- Large organizations make larger absolute investments with longer payback expectations
- Small organizations focus on low-cost solutions with immediate returns

- Mid-size organizations carefully balance investment with clear value demonstration requirements

Geographic and Cultural Considerations

While our survey data provides limited geographic breakdown, industry and size patterns suggest important geographic implications:

Technology hub advantages

Organizations in technology centers (Silicon Valley, Seattle, Boston, Austin, etc.) demonstrate:

- Earlier adoption and more sophisticated implementations
- Greater access to AI talent and specialized partners
- More aggressive competitive positioning using AI capabilities
- Higher comfort with experimental and cutting-edge implementations

Traditional business centers

Organizations in established business centers show:

- More conservative but systematic implementation approaches
- Focus on proven applications with clear ROI
- Greater emphasis on integration with existing business processes
- Careful attention to regulatory and compliance requirements

Regional variations

- Rural and smaller market organizations face greater challenges accessing AI talent and specialized services

- International organizations must navigate varying regulatory environments and cultural expectations
- Regional competitive dynamics influence adoption urgency and implementation approaches

Strategic Implications of Contextual Variations

Understanding these industry and size-based variations provides several strategic insights:

Implementation strategy customization: Organizations should develop implementation approaches appropriate to their industry context, regulatory environment, and resource capabilities rather than adopting generic strategies.

Competitive benchmarking: Organizations benefit from understanding adoption patterns within their specific industry and size category to set appropriate competitive expectations and timelines.

Partnership strategy development: Industry and size characteristics suggest optimal partnership approaches, with smaller organizations requiring different support than large enterprises.

Investment planning: Resource allocation strategies should reflect industry competitive dynamics and organizational size constraints while building toward long-term strategic positioning.

These contextual variations demonstrate that while agentic AI adoption follows certain universal patterns, successful implementation requires careful attention to industry-specific requirements, organizational capabilities, and competitive dynamics that vary significantly across different business contexts.

Perhaps the most striking finding from our survey relates to organizational expectations about the transformative impact of agentic AI. These expectations provide crucial insights into how business leaders anticipate market evolution and competitive dynamics over the next five years.

Overwhelming Transformation Confidence

When asked to rate their agreement with the statement "Agentic AI will fundamentally transform our industry within the next five years," respondents demonstrated remarkable confidence:

- **Strongly Agree (Rating 5)**: 48.5% (194 organizations)
- **Agree (Rating 4)**: 31.3% (125 organizations)
- **Neutral (Rating 3)**: 13.8% (55 organizations)
- **Disagree (Rating 2)**: 3.8% (15 organizations)
- **Strongly Disagree (Rating 1)**: 2.7% (11 organizations)

The headline insight: 79.8% of organizations expect industry transformation within five years, with nearly half (48.5%) expressing strong confidence.

Only 6.5% are skeptical, suggesting a broad consensus across industries and organizational sizes about agentic AI's transformative impact.

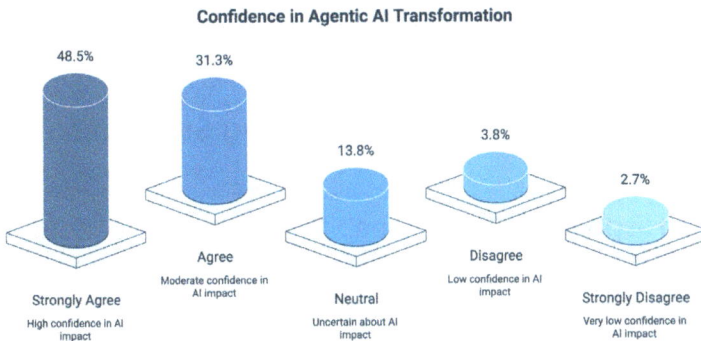

Confidence in Agentic AI Transformation

48.5% — Strongly Agree — High confidence in AI impact
31.3% — Agree — Moderate confidence in AI impact
13.8% — Neutral — Uncertain about AI impact
3.8% — Disagree — Low confidence in AI impact
2.7% — Strongly Disagree — Very low confidence in AI impact

Competitive Response Patterns: Urgency Without Panic

Understanding how organizations would respond to major competitive AI implementations reveals strategic thinking about competitive dynamics:

Competitive response distribution

- **Accelerate existing agentic AI initiatives**: 31.7% (127 organizations)
- **Maintain current pace**: 26.2% (105 organizations)
- **Launch new agentic AI initiatives**: 22.5% (90 organizations)
- **Conduct competitive analysis first**: 12.5% (50 organizations)
- **Wait to see results**: 7.0% (28 organizations)

Strategic insights

- 54.2% would accelerate or launch new initiatives in response to competitive pressure
- 26.2% would maintain the current pace, suggesting confidence in existing strategies
- Only 7.0% would wait for results before responding, indicating minimal complacency

This response pattern suggests organizations recognize competitive urgency while maintaining strategic discipline rather than panic-driven reactions.

Future Capability Priorities: Building Sustainable Advantage

Survey responses about capabilities needed for future success reveal strategic priorities for sustainable competitive advantage:

Critical capability areas (rated as "Very Important" or "Critically Important")

- **Integration with legacy systems**: Priority for operational effectiveness
- **Advanced analytics capabilities**: Foundation for data-driven decision making
- **Specialized AI talent acquisition**: Recognition of human capital constraints
- **Cross-functional collaboration**: Emphasis on organizational adaptation
- **Agile implementation methodologies**: Focus on execution speed and flexibility
- **Executive-level AI literacy**: Leadership development priority

These priorities suggest organizations understand that sustainable advantage requires comprehensive capability development rather than technology implementation alone.

Investment Trajectory Indicators

Current budget allocation patterns provide insights into future investment trends:

Technology budget allocation to agentic AI

- **0-5%**: 32.1% of organizations (early stage allocation)
- **6-15%**: 41.2% of organizations (substantial commitment)
- **16-25%**: 18.3% of organizations (significant investment)
- **26%+**: 8.4% of organizations (major strategic focus)

Investment insights

- 67.9% allocate 6% or more of technology budgets to agentic AI
- 26.7% allocate 16% or more, indicating substantial strategic commitment

- Investment distribution suggests growing rather than stable allocation patterns

Transformation Timeline Implications

The five-year transformation expectation creates several important strategic implications:

Acceleration imperative: Accelerating implementation beyond the current pace to capture competitive advantages before they become industry standards.

Window of opportunity: Using the limited window available to establish competitive advantages through superior agentic AI implementation before widespread adoption equalizes basic capabilities.

Ecosystem development: Driving rapid development of supporting ecosystems, including talent, services, standards, and regulations.

Market structure evolution: Understanding that the coming transformation may challenge current market leaders as new entrants become better positioned to leverage agentic AI capabilities.

Strategic Planning Implications

High transformation confidence creates several strategic planning requirements:

Scenario planning necessity: Developing multiple scenarios for industry evolution rather than single-point predictions, given the high confidence in transformation combined with uncertainty about specific forms.

Capability building urgency: Building immediate capabilities in areas identified as critical for future success, rather than gradual development approaches.

Competitive positioning review: Reassessing competitive positioning assumptions based on potential industry transformation rather than extrapolating current competitive dynamics.

Partnership strategy evolution: Preparing for new partnership strategies as ecosystem participants, customer requirements, and competitive dynamics evolve.

Risk and Opportunity Balance

The survey findings suggest organizations should balance transformation opportunities with implementation risks:

Opportunity maximization: Organizations with high confidence in transformation gain outsized rewards if they execute well. Building capabilities early positions them to lead as transformation accelerates, while strategic moves during transformation can create sustainable competitive advantages.

Risk management: Organizations need flexible approaches to handle uncertainties during transformation and adapt to shifting markets. Overconfidence in specific trajectories risks poor investment choices, while fast-moving competitors demand ongoing strategy realignment.

Preparation Priorities Based on Expectations

The combination of high transformation confidence with current implementation challenges suggests several preparation priorities:

Technical foundation: Organizations should invest in data infrastructure, integration capabilities, and technical skills that enable rapid scaling as transformation accelerates.

Organizational adaptation: Culture, processes, and governance frameworks should evolve to support increased agentic AI integration before competitive pressure intensifies.

Talent strategy: Human capital strategies should anticipate changing skill requirements and competitive talent markets rather than addressing current gaps alone.

Strategic flexibility: Implementation approaches should preserve flexibility to adapt as transformation unfolds rather than locking in specific technical or strategic directions.

The overwhelming confidence in industry transformation, combined with sophisticated strategic thinking about competitive implications, suggests organizations are preparing for disruptive rather than incremental change.

Success will depend on balancing urgency with strategic discipline, building comprehensive capabilities while maintaining adaptive flexibility, and executing effectively during a period of rapid market evolution.

Conclusion: A Market in Strategic Transformation

Our comprehensive survey of 442 business professionals reveals an agentic AI adoption landscape that challenges common assumptions about market maturity and strategic sophistication.

Rather than finding organizations in early experimental phases, we discovered a market that has moved decisively beyond exploration into active implementation, with sophisticated strategic thinking about competitive implications and transformation potential.

Key Market Reality: Beyond Early Adoption

The most significant finding challenges conventional wisdom about agentic AI market maturity. With 89.1% of organizations already exploring or implementing agentic AI, and 49.1% having active production deployments, the market has reached a tipping point where implementation experience rather than theoretical exploration dominates organizational thinking.

This maturity level creates new competitive dynamics where advantage flows to organizations that execute most effectively rather than those that simply adopt first.

The implication for organizational strategy is profound: the question has shifted from "whether" to implement agentic AI to "how" to implement it most successfully.

Strategic Sophistication: Competitive Transformation Focus

Organizations demonstrate remarkable strategic sophistication in their approach to agentic AI. Rather than viewing these technologies primarily as cost-reduction tools (only 18.5%), they see competitive advantage (26.8%), transformative opportunity (24.8%), or competitive necessity (19.5%). This advanced strategic thinking suggests that agentic AI has become central to competitive strategy.

The 79.8% of organizations expecting an industry transformation within five years indicates a broad consensus about change magnitude while creating urgency for capability building and strategic positioning.

Implementation Patterns: Function-by-Function Evolution

Clear patterns emerge in implementation approaches, with IT and security (53.9%) and customer service (46.8%) leading adoption, followed by operations, sales and marketing, and finance in the 30-40% range. Legal and compliance lag significantly (13.1%), reflecting appropriate caution in high-stakes applications.

These patterns reveal a risk-calibrated approach where organizations begin with lower-risk, higher-value applications before expanding to more complex domains. This suggests a maturation pathway that other organizations can follow, using early adopters' experiences to guide their own implementation sequencing.

The Execution Gap: Technical and Human Challenges

Despite clear value recognition, only 9.9% cite unclear ROI as a barrier; organizations face substantial execution challenges led by technical integration complexity (48.5%), skills/talent gaps (42.1%), and data quality issues (40.6%). This pattern **reveals a market where strategic vision exceeds current execution capabilities**, creating opportunities for organizations that can solve implementation challenges effectively.

The dominance of technical and human capital barriers over business case concerns suggests the market has moved beyond value demonstration to capability-building phases. Organizations should prioritize technical integration capabilities, skills development, and data infrastructure as prerequisites for competitive advantage.

Industry and Size Variations: Context Matters

Implementation patterns vary significantly across industries and organizational sizes, with technology companies leading adoption while regulated industries proceed more cautiously. Large enterprises achieve more comprehensive implementations while smaller organizations focus on targeted, high-impact applications.

These variations demonstrate that successful agentic AI adoption requires contextual strategies rather than generic approaches. Organizations benefit from understanding adoption patterns within their specific industry and size category to set appropriate competitive expectations and develop realistic implementation timelines.

Strategic Implications: Preparing for Accelerating Change

The survey findings suggest several crucial strategic implications for organizational leaders:

Competitive urgency without panic: While 54.2% of organizations would accelerate AI initiatives in response to competitive pressure, the measured response patterns suggest strategic discipline rather than panic-driven reactions. Organizations should maintain focus on execution quality while recognizing competitive timing pressures.

Capability building priority: With transformation expected within five years, organizations must immediately begin building comprehensive capabilities in technical integration, talent development, and organizational adaptation rather than waiting for technology to mature further.

Strategic positioning window: The combination of high adoption rates with execution challenges creates a limited window for establishing competitive advantages through superior implementation before basic capabilities become commoditized across industries.

Ecosystem evolution acceleration: High transformation confidence will drive rapid development of supporting ecosystems, including talent pools, service providers, regulatory frameworks, and industry standards, requiring proactive engagement rather than reactive adaptation.

Forward-Looking Guidance: Strategic Recommendations

Based on our survey findings, organizations should consider several strategic recommendations:

Immediate actions

1. **Assess implementation readiness**: Conduct a comprehensive evaluation of technical integration capabilities, talent availability, and data infrastructure quality.
2. **Develop phased implementation plans**: Create realistic timelines that address execution challenges while maintaining competitive positioning.

3. **Invest in capability building**: Prioritize technical integration expertise, AI talent development, and organizational change management capabilities.

Medium-term strategy

1. **Build strategic flexibility**: Develop implementation approaches that can adapt as technology capabilities and competitive dynamics evolve.
2. **Create learning systems**: Establish mechanisms to capture implementation insights and accelerate organizational learning across initiatives.
3. **Develop partnership strategies**: Form relationships with technology providers, system integrators, and consulting firms that can address capability gaps.

Long-term positioning

1. **Prepare for transformation**: Build organizational foundations that can support business model evolution as transformation accelerates.
2. **Maintain competitive focus**: Ensure agentic AI investments align with broader strategic positioning rather than pursuing implementation for its own sake.
3. **Develop market leadership**: Position organization to influence industry standards, practices, and ecosystem development rather than simply following market evolution.

The Path Forward: From Implementation to Transformation

The data from our survey paints a picture of a business environment in the midst of fundamental change. What emerges is not the cautious, experimental approach to agentic AI that many might expect, but rather a confident, strategic embrace of technologies that organizations believe will reshape entire industries within five years or less.

This shift from "whether" to "how" marks an inflection point where competitive advantage increasingly depends on execution excellence rather than early adoption timing.

The path forward requires organizations to navigate three simultaneous challenges: solving immediate implementation barriers while building long-term competitive capabilities, maintaining strategic flexibility as transformation accelerates, and developing comprehensive organizational readiness that spans technical infrastructure, human capital, and change management.

Organizations that master this multi-dimensional challenge will not simply adopt agentic AI—they will use it to define new standards of performance, customer experience, and market leadership in their industries.

Perhaps most significantly, our findings suggest that the window for establishing competitive advantage through superior agentic AI implementation may be narrower than many assume.

With transformation expectations concentrated in a five-year timeframe and adoption rates already reaching critical mass, the organizations that act decisively now—guided by the patterns and insights revealed in this research—will be best positioned to lead rather than follow in the emerging digital workforce economy.

CHAPTER 11:
FUTURE TRAJECTORIES OF THE DIGITAL WORKFORCE

TL;DR:

- Digital worker capabilities will continue to evolve rapidly, with near-term enhancements in reasoning, multimodal understanding, tool use, and context management enabling increasingly complex applications.

- Longer-term developments may include more autonomous learning, advanced collaboration capabilities, enhanced creativity, and specialized domain expertise that could transform how organizations operate.

- The economic and social implications of digital workforces will be significant, requiring thoughtful approaches to workforce transition, skill development, and broader societal adaptation.

- Organizations face important strategic positioning decisions regarding implementation timing, capability development priorities, partnership approaches, and competitive differentiation strategies.

- Building adaptive capacity through organizational flexibility, evolved talent strategies, sophisticated investment prioritization, and scenario-based planning will enable organizations to navigate uncertainty while capturing emerging opportunities.

Developing a Future-looking Strategy

Understanding emerging developments and likely evolution paths enables organizations to develop forward-looking strategies that anticipate rather than merely react to technological and market changes. These insights help leaders make implementation decisions today that align with longer-term digital workforce evolution.

Technological Evolution: Anticipating Future Capabilities

The capabilities of digital workers continue to advance rapidly, with several clear trajectories that will significantly expand their potential application domains and value creation.

Near-Term Development Roadmaps

Over the next 12-24 months, several capability enhancements will reach mainstream implementation maturity, creating new opportunities for digital workforce applications.

Emerging capability areas

1. **Enhanced reasoning and problem-solving**

- More sophisticated multi-step reasoning
- Improved handling of complex logic and constraints
- Enhanced causal reasoning capabilities
- Better management of uncertainty and ambiguity
- More robust error detection and correction

2. **Expanded multimodal understanding**

- More seamless integration across text, images, and data
- Enhanced ability to extract meaning from visual content
- Better understanding of spatial relationships

- Improved processing of charts, diagrams, and technical images
- More natural integration of information across modalities

3. **Advanced tool utilization**

- More sophisticated use of specialized software
- Enhanced ability to orchestrate multiple tools
- Better adaptation to new tool interfaces and capabilities
- Improved error handling during tool interactions
- More flexible integration with enterprise systems

4. **Richer context management**

- Longer and more nuanced conversation handling
- Better maintenance of relevant context over time
- Enhanced user-specific personalization
- Improved organizational memory and knowledge application
- More sophisticated project and process awareness

Implementation implications

Advancing Digital Worker Potential

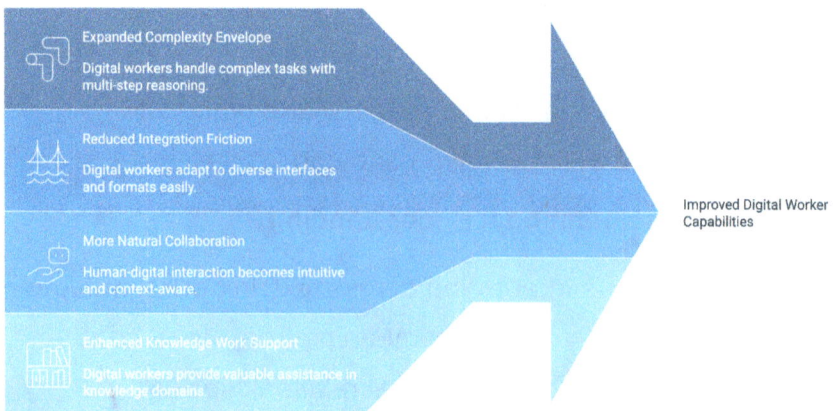

Expanded Complexity Envelope
Digital workers handle complex tasks with multi-step reasoning.

Reduced Integration Friction
Digital workers adapt to diverse interfaces and formats easily.

More Natural Collaboration
Human-digital interaction becomes intuitive and context-aware.

Enhanced Knowledge Work Support
Digital workers provide valuable assistance in knowledge domains.

Improved Digital Worker Capabilities

These digital worker capability enhancements will enable several important implementation evolutions:

Expanded complexity envelope: Handling increasingly complex tasks that require multi-step reasoning, contextual judgment, and integration across information types.

Reduced integration friction: Connecting with existing systems will become more straightforward as they develop better capabilities to adapt to diverse interfaces and formats.

More natural collaboration: Interacting with humans will become more intuitive as digital workers better maintain context, understand nuance, and adapt to individual collaboration styles.

Enhanced knowledge work support: Providing more valuable assistance in knowledge-intensive domains as their reasoning, context management, and multimodal understanding improve.

Organizations should monitor these developments closely and prepare implementation strategies to leverage these emerging capabilities as they mature.

Medium-Term Capability Expansion Possibilities

Looking further ahead (3-5 years), more capability expansions will likely emerge, potentially transforming how digital workers operate and create value.

Transformative capability directions

1. **Autonomous learning and adaptation**

- Self-directed capability enhancement based on experience
- Proactive pattern recognition and improvement suggestion
- Context-specific optimization without explicit training
- Adaptation to shifting environments and requirements
- Continuous performance enhancement with minimal supervision

2. **Advanced collaboration capabilities**

- Sophisticated team integration across human and digital workers
- Enhanced understanding of organizational dynamics
- Nuanced communication tailored to different stakeholders
- Initiative-taking within appropriate boundaries
- Improved emotional intelligence and relationship management

3. **Enhanced creative capabilities**

- More sophisticated ideation and concept generation
- Better evaluation of novel solutions and approaches
- Enhanced ability to combine concepts across domains
- More nuanced aesthetic and design judgment
- Improved capability to iterate and refine creative work

4. **Specialized domain expertise**

- Deep knowledge in complex professional domains
- Integration of theoretical knowledge with practical application
- Reasoning that incorporates domain-specific constraints
- Ability to explain advanced concepts at different levels
- Contextually appropriate reference to specialized knowledge

AI capabilities range from basic to highly advanced.

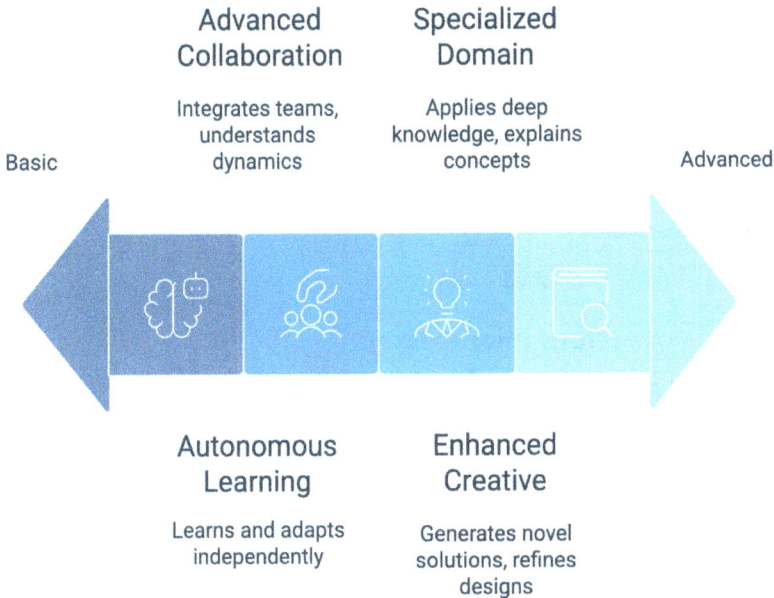

Advanced
Collaboration

Specialized
Domain

Integrates teams,
understands
dynamics

Applies deep
knowledge, explains
concepts

Basic

Advanced

Autonomous
Learning

Enhanced
Creative

Learns and adapts
independently

Generates novel
solutions, refines
designs

Organizational transformation implications

These medium-term developments will enable more changes in how organizations leverage digital workforces:

Self-improving systems: Increasing enhancement in their own capabilities to reduce the need for explicit retraining and enable continuous performance improvement.

Autonomous workflow orchestration: Coordinating more complex workflows beyond executing individual tasks to manage both task sequencing and resource allocation.

Strategic support expansion: Providing more valuable input for strategic decisions by synthesizing information, generating alternatives, and evaluating potential outcomes.

Professional domain transformation: Increasing integration of digital workers into knowledge-intensive professions as core team members rather than merely support tools, changing workflow, and value creation.

Organizations that anticipate these developments make architectural and strategic decisions today that position them to efficiently incorporate these capabilities as they mature.

Long-Term Transformative Potential

Beyond the five-year horizon, several potentially transformative developments may reshape how organizations conceptualize and utilize digital workforces.

Potential transformation vectors

1. **Emergent intelligence patterns**

- Capabilities exceeding the sum of the designed components
- Novel problem-solving approaches not explicitly programmed
- Unexpected pattern recognition and insight generation
- Self-directed capability expansion into new domains
- Sophisticated adaptation to previously unencountered situations

2. **Seamless human-digital integration**

- Boundary blurring between human and digital work
- Intuitive collaboration requiring minimal explicit direction
- Predictive support anticipating needs before articulation
- Personalized adaptation to individual working styles
- Dynamic capability expansion to complement human strengths

3. **Expanded sensory and interaction capabilities**

- Integration with physical world sensing technologies

- Enhanced spatial and environmental awareness
- Multimodal input and output across diverse channels
- More natural and intuitive interaction mechanisms
- Expanded ability to process non-textual information

4. **Collective intelligence emergence**

- Network effects across digital worker instances
- Distributed learning and capability sharing
- Coordinated problem-solving across specializations
- Ecosystem-level adaptation and optimization
- Collaborative intelligence beyond individual capabilities

Transformation Vectors Comparison

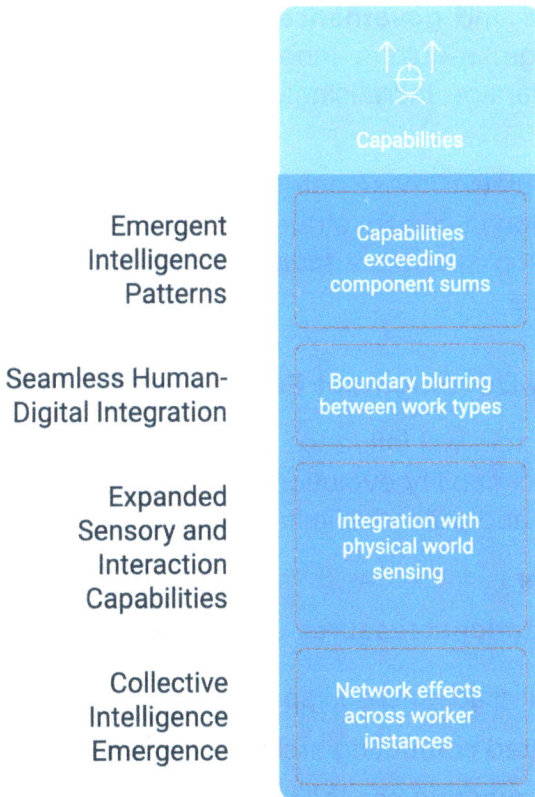

	Capabilities
Emergent Intelligence Patterns	Capabilities exceeding component sums
Seamless Human-Digital Integration	Boundary blurring between work types
Expanded Sensory and Interaction Capabilities	Integration with physical world sensing
Collective Intelligence Emergence	Network effects across worker instances

Societal and organizational implications

These long-term possibilities suggest several profound potential shifts:

Work redefinition: The nature of many forms of work may be reimagined, focusing human contribution on areas of unique value while transforming or eliminating many current roles.

Organizational structure evolution: Traditional hierarchies and functional divisions may give way to more fluid, capability-centered structures enabled by digital workforce capabilities.

Value creation transformation: How organizations create and capture value may change, with emphasis shifting from human resource management to digital-human ecosystem orchestration.

Novel risks and governance challenges: New forms of risk may emerge, requiring innovative governance approaches that can manage increasingly autonomous and capable systems.

While speculative, these potential long-term directions warrant consideration in current strategic thinking to ensure organizations develop adaptable foundations that evolve as capabilities advance.

Capability Development Enablers and Constraints

Multiple factors will influence the pace and direction of digital workforce capability evolution, creating both acceleration opportunities and potential limitations.

Key enablers

1. **Computational resource expansion**

- Continued advances in specialized AI hardware
- Distributed computing architecture improvements
- Edge computing capability enhancement

- Memory technology evolution
- Energy efficiency improvements

2. **Data availability and quality**

- Improved high-quality training datasets
- Improved data collection and annotation techniques
- Enhanced synthetic data generation
- Better domain-specific data availability
- Advanced data quality assurance methods

3. **Algorithm and architecture innovation**

- Novel neural network architectures
- Improved training methodologies
- Enhanced optimization techniques
- Better transfer learning approaches
- Advanced performance scaling methods

Key Enablers

Computational Resources — Advances in hardware, architecture, and memory.

Data Availability — Improved datasets, collection, and synthetic data.

Algorithm Innovation — Novel architectures, training, and optimization techniques.

Significant constraints

1. **Technical limitations**

- Reasoning consistency challenges
- Physical world interaction constraints
- Resource requirements for the largest models
- Security and robustness challenges
- System integration complexity

2. **Regulatory and ethical boundaries**

- Emerging regulatory frameworks
- Safety and oversight requirements
- Privacy protection obligations
- Transparency and explainability expectations
- Ethical implementation standards

3. **Organizational adoption barriers**

- Implementation complexity
- Skills and capability gaps
- Cultural resistance
- Legacy system integration challenges
- Risk management concerns

Understanding these enablers and constraints helps organizations develop realistic expectations about capability evolution while identifying opportunities to accelerate value capture where possible.

Economic and Social Implications: Broader Impacts and Considerations

The expansion of digital workforces will create significant economic and social impacts that extend far beyond immediate implementation considerations. Understanding these broader implications is essential for sustainable strategy development.

Labor Market Evolution Scenarios

The specific nature of the impacts of digital workforce expansion remains subject to considerable debate and uncertainty.

Potential evolution patterns

1. **Transition and displacement effects**

- Routine cognitive task automation
- Administrative and support role transformation
- Middle-skill job evolution or elimination
- Particular vulnerability in data processing and regular analysis domains
- Disproportionate impact on early-career entry roles

2. **Job creation and enhancement**

- New roles orchestrating digital-human collaboration
- Expanded positions in AI management and oversight
- Growth in experience design and system integration
- Increased demand for uniquely human capabilities
- New work categories enabled by digital capabilities

3. **Skill demand shifts**

- Declining value of routine information processing
- Increasing premium on creative problem-solving

- Growing importance of emotional intelligence
- Rising value of technical-business translation abilities
- Enhanced demand for systems thinking and integration skills

Probable impact distribution

The effects of digital workforce expansion will likely be unevenly distributed:

Industry variation: Knowledge-intensive sectors like financial services, legal, and professional services may see more significant transformation than physically-oriented industries.

Geographic differences: Regions with higher concentrations of routine cognitive work may experience greater disruption than those with more diverse economies.

Temporal variation: Impacts will unfold over time rather than appearing simultaneously, with some effects becoming apparent within 2-3 years while others emerge over decades.

Skill-level divergence: Workers with high adaptability, specialized expertise, or distinctly human capabilities may benefit from complementary effects, while those focused on routine tasks face substitution.

Organizations should consider these potential patterns in workforce planning, skill development, and social responsibility strategies while recognizing the significant uncertainty in specific predictions.

Skill Development Imperatives

As digital workforces evolve, the skills required for human workers to create value will shift significantly, creating both challenges and opportunities for individuals and organizations.

Critical emerging skill categories

1. **Human-digital collaboration skills**

- Effective direction and instruction provision
- Output evaluation and quality assessment
- Appropriate trust calibration
- Strategic task allocation
- Process design incorporating digital capabilities

2. **Complex reasoning and judgment**

- Problem framing and context setting
- Ethical evaluation and value alignment
- Managing uncertainty
- Systems thinking and interconnection recognition
- Balancing competing priorities and considerations

3. **Creative and innovative thinking**

- Novel concept generation
- Cross-domain connection identification
- Challenging established approaches
- Effective ideation and experimentation
- Adapting to emergent possibilities

4. **Social and emotional intelligence**

- Relationship building and maintenance
- Emotional understanding and empathy
- Cultural sensitivity and awareness
- Conflict resolution and negotiation
- Team building and motivation

Emerging Skill Categories

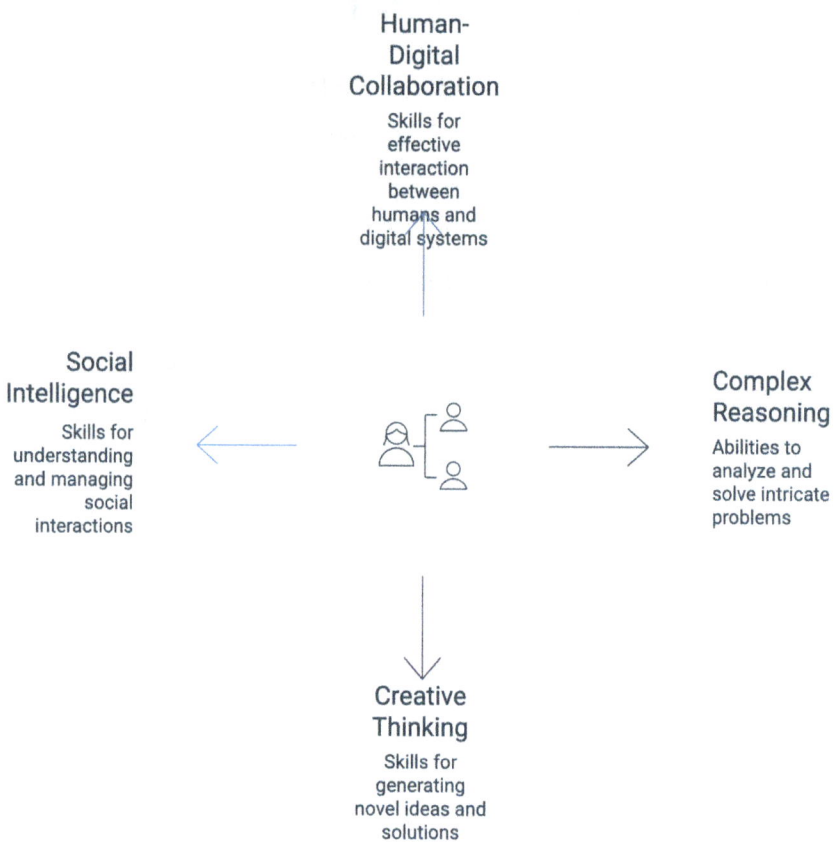

Human-Digital Collaboration

Skills for effective interaction between humans and digital systems

Social Intelligence

Skills for understanding and managing social interactions

Complex Reasoning

Abilities to analyze and solve intricate problems

Creative Thinking

Skills for generating novel ideas and solutions

Organizational response requirements

Organizations must develop systematic approaches to address these skill evolution needs:

Learning ecosystem development: Creating comprehensive learning environments that combine formal training, experiential learning, and continuous development.

Career path evolution: Redesigning career progression to emphasize continuous skill expansion and adaptation rather than fixed role mastery.

Education partnership: Collaborating with educational institutions to align curricula with emerging skill requirements.

Inclusive transition support: Ensuring development opportunities reach all workforce segments rather than concentrating only on already-advantaged groups.

Hiring and selection evolution: Adapting recruitment to emphasize adaptability, creativity, and collaboration potential alongside technical capabilities.

Organizations that proactively address these skill development imperatives will create significant competitive advantages through superior human-digital workforce integration.

Distributional Effects and Policy Considerations

The economic benefits and challenges of digital workforce expansion may be unevenly distributed across society, raising important policy questions organizations should consider.

Potential distributional challenges

1. **Labor share of income**

- Balance shift between labor and capital returns
- Concentration of productivity benefits
- Wage pressure on automatable roles
- Market power enhancement for technology leaders
- Geographic concentration of economic returns

2. **Opportunity access variations**

- Digital divide expansion risks
- Educational preparation disparities
- Retraining access differences
- Regional economic divergence
- Generational adaptation variation

3. **Transition support gaps**

- Uneven safety net coverage
- Retraining program adequacy
- Mid-career transition challenges
- Industry-specific displacement concentration
- Temporary versus permanent impact management

Organizational consideration areas

Forward-thinking organizations should consider several policy-relevant dimensions:

Transition support approaches: How to provide effective assistance for workers whose roles evolve or are eliminated due to digital workforce implementation.

Value distribution mechanisms: How to ensure productivity and profit gains benefit a broad stakeholder base rather than concentrating narrowly.

Community impact management: How to address potential effects on communities where organizations operate, particularly in regions with concentrated employment.

Policy engagement strategy: How to constructively participate in developing appropriate policy frameworks that balance innovation and inclusion.

Ethical implementation standards: How to establish and maintain implementation approaches that align with organizational values and societal expectations.

Organizations that thoughtfully address these considerations will build more sustainable implementation approaches and stronger stakeholder relationships than those focusing exclusively on immediate business benefits.

Societal Adaptation Requirements

Beyond economic impacts, digital workforce expansion may create broader societal adaptation requirements that warrant consideration in organizational strategy.

Social system evolution needs

1. **Education system transformation**

- Curriculum evolution for complementary skills
- Lifelong learning infrastructure development
- Education financing model adaptation
- Educational credential evolution
- Human-digital learning integration

2. **Social support mechanism adaptation**

- Career transition support enhancement
- Economic security approach evolution
- Regional economic development strategies
- Community resilience building
- Inclusive opportunity creation

3. **Social contract reconsideration**

- Work and identity relationship evolution
- Value creation recognition frameworks
- Contribution beyond employment
- Technological benefit distribution
- Responsibility allocation for transition

Navigating Social System Evolution

Education System Transformation
Adapting education for future skills

Social Support Mechanism Adaptation
Enhancing support for career and economic security

Social Contract Reconsideration
Rethinking work, identity, and responsibility

Organizational leadership opportunities

Organizations can play constructive roles in societal adaptation through several approaches:

Multi-stakeholder collaboration: Participating in cross-sector initiatives addressing broader adaptation needs beyond immediate business interests.

Responsible implementation leadership: Developing and sharing approaches that maximize positive impacts while minimizing disruption.

Community engagement: Working with local communities to build adaptive capacity and create inclusive opportunities.

Public discourse contribution: Adding constructive perspective to public conversations about technological change and social adaptation.

Policy development support: Providing practical insight to policymakers about effective approaches to managing technological transition.

Organizations that engage thoughtfully with these broader societal considerations typically build stronger stakeholder relationships, enhance reputation, and contribute to creating environments where technological innovation can thrive sustainably.

Competitive Landscape: Strategic Positioning for Future Advantage

As digital workforces become increasingly central to organizational performance, strategic positioning in this domain will significantly influence overall competitive standing. Understanding emerging patterns informs more effective long-term strategies.

First-Mover Advantages vs. Fast-Follower Strategies

Organizations face important timing decisions regarding digital workforce implementation, requiring careful consideration of potential advantages and risks in different strategic approaches.

First-mover potential advantages

1. **Data and learning accumulation**

- Earlier access to implementation experience
- Larger datasets for optimization and learning
- More extensive feedback for refinement
- Longer capability maturation timeline
- Greater institutional knowledge development

2. **Talent acquisition and development**

- Earlier access to limited specialist talent

- More extensive skills development opportunity
- Stronger employer brand in critical disciplines
- Deeper expertise accumulation
- More mature organizational capabilities

3. **Strategic positioning benefits**

- Opportunity to set industry standards
- Potential to establish platform positions
- Client and partner relationship advantages
- Ability to influence regulatory development
- Opportunity to shape market expectations

Fast-follower potential advantages

1. **Learning from pioneer experiences**

- Lessons from early implementation mistakes
- Access to established best practices
- Building on proven use cases
- Reduced experimentation costs
- Clearer ROI expectations

2. **Technology maturation benefits**

- Access to more developed solutions
- Lower implementation costs over time
- Reduced integration complexity
- More robust security and governance frameworks
- Established implementation methodologies

3. **Market timing optimization**

- Better alignment with customer readiness
- Reduced change management challenges

- Clearer competitive differentiation opportunities
- More developed ecosystem of partners
- Better regulatory clarity

Strategic approach considerations

Most organizations will benefit from nuanced timing strategies that combine elements of both approaches:

Portfolio balancing: Pursuing selective first-mover positions in strategically crucial domains while taking fast-follower approaches in others.

Capability building focus: Developing core organizational capabilities early, even if full-scale implementation follows a more measured timeline.

Strategic experimentation: Conducting focused pilots to build experience while deferring large-scale deployment until greater maturity.

Ecosystem positioning: Engaging early in emerging digital workforce ecosystems while maintaining implementation flexibility.

Risk-calibrated timing: Accelerating in lower-risk domains while taking more measured approaches where potential negative consequences are greater.

Which strategic approach should be prioritized for digital workforce implementation?

Ecosystem Positioning

Engage early in digital ecosystems for flexibility.

Capability Building

Develop core capabilities early for future implementation.

Risk-Calibrated Timing

Accelerate in low-risk areas and proceed cautiously in high-risk ones.

Strategic Experimentation

Conduct pilots to gain experience before large-scale deployment.

Portfolio Balancing

Pursue first-mover positions in crucial domains while following in others.

Organizations should develop explicit digital workforce timing strategies based on their specific competitive context, risk profile, and strategic priorities rather than defaulting to either aggressive pursuit or excessive caution.

Core Competency Development Priorities

Regardless of implementation timing, organizations must develop several core competencies to successfully leverage digital workforces for competitive advantage.

Strategic capability categories

1. **Digital workforce architecture**

- System component integration design
- Human-digital workflow orchestration
- Enterprise system connection frameworks
- Data flow and knowledge management
- Scalable and adaptable foundation creation

2. **Implementation and change excellence**

- Use case identification and prioritization
- Value capture methodology
- Organizational adaptation management
- User adoption and experience optimization
- Continuous improvement processes

3. **Risk and governance management**

- Appropriate control framework development
- Ethical implementation approaches
- Regulatory compliance methodologies
- Security and resilience capabilities
- Responsible innovation practices

4. **Strategic workforce integration**

- Human-digital team design
- Capability-centered role definition
- Career path and development framework creation
- Performance management adaptation
- Cultural evolution management

Development approach elements

Organizations can accelerate competency development through several approaches:

Centers of excellence: Creating dedicated teams focused on building and disseminating critical capabilities across the organization.

Strategic partnerships: Forming relationships with technology providers, consulting firms, and implementation specialists to access specialized expertise.

Targeted acquisition: Acquiring organizations selectively with established capabilities in critical domains.

Talent investment: Recruiting, developing, and retaining specialists in high-priority capability areas.

Experiential learning: Designing implementation approaches that explicitly incorporate learning objectives alongside business outcomes.

Organizations that develop these competencies create foundation capabilities that enable effective execution regardless of specific implementation timing decisions.

Strategic Partnership Considerations

Few organizations are able to develop all required digital workforce capabilities internally, making strategic partnership decisions increasingly important for competitive positioning.

Partnership domain framework

1. **Foundation technology relationships**

- AI model and platform providers
- Integration infrastructure partners
- Development environment providers
- Specialized capability licensors
- Computing infrastructure suppliers

2. **Implementation support ecosystems**

- Design and development partners
- Change management specialists
- Training and enablement providers
- Industry-specific solution developers
- Managed service operators

3. **Domain knowledge partnerships**

- Industry specialist collaborations
- Subject matter expert relationships
- Content and knowledge providers
- Training data sources
- Specialized capability developers

Partnership Framework Structure

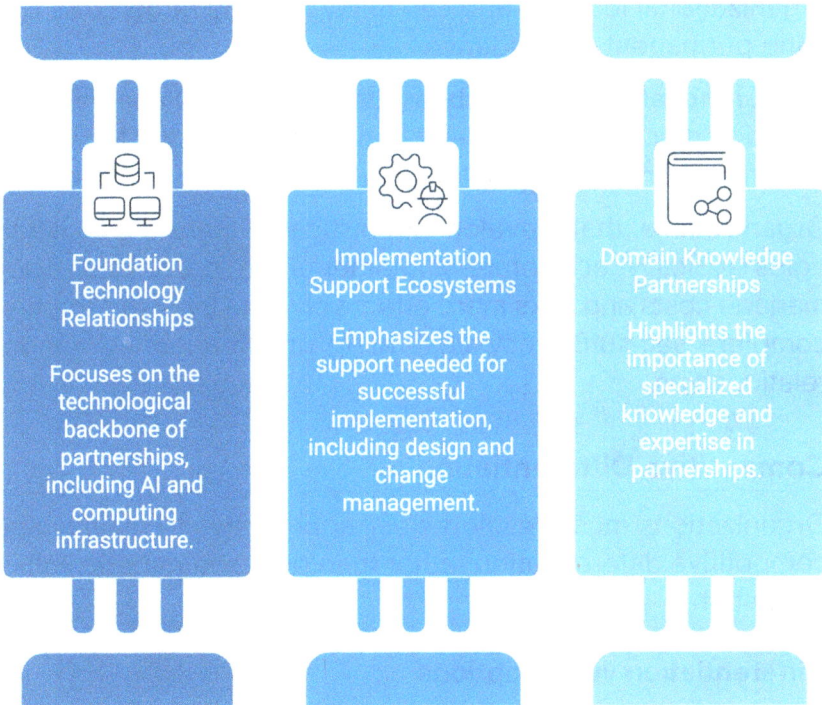

Foundation Technology Relationships

Focuses on the technological backbone of partnerships, including AI and computing infrastructure.

Implementation Support Ecosystems

Emphasizes the support needed for successful implementation, including design and change management.

Domain Knowledge Partnerships

Highlights the importance of specialized knowledge and expertise in partnerships.

Partnership strategy elements

Organizations should consider several dimensions in partnership strategy development:

Make vs. buy vs. partner decisions: Determining which capabilities to develop internally versus access through partnerships based on strategic importance, differentiation potential, and resource requirements.

Ecosystem position optimization: Establishing positions within emerging partner ecosystems that maximize influence and access while maintaining flexibility.

Dependency risk management: Balancing partnership benefits against potential dependency risks through appropriate contract structures, diversification, and internal capability development.

Knowledge transfer focus: Designing partnerships to facilitate organizational learning and capability building rather than creating permanent dependence.

Joint innovation approaches: Establishing collaborative innovation models that leverage combined strengths for mutual benefit rather than transactional relationships.

Organizations that develop sophisticated partnership strategies typically accelerate implementation processes. They manage costs and risks more effectively than those attempting complete self-sufficiency or forming undifferentiated vendor relationships.

Competitive Differentiation

Organizations must develop clear approaches to maintaining competitive differentiation rather than pursuing generic implementation.

Differentiation vector options

1. **Specialized capability development**

- Industry-specific digital worker creation
- Unique function optimization
- Proprietary data advantage leveraging
- Custom integration with strategic systems

- Organization-specific knowledge incorporation

2. **Experience and interface innovation**

- Superior human-digital collaboration design
- Unique user experience creation
- Distinctive customer interaction models
- Innovative employee augmentation approaches
- Novel workflow integration techniques

3. **Operating model transformation**

- Process redesign
- Organizational structure innovation
- Novel human-digital role definition
- Ecosystem orchestration approaches
- Business model reinvention

4. **Culture and talent integration**

- Distinctive human-digital teaming
- Superior adoption and utilization
- Enhanced learning and adaptation
- Accelerated improvement cycles
- Unique collaboration patterns

Differentiation Vector Cycle

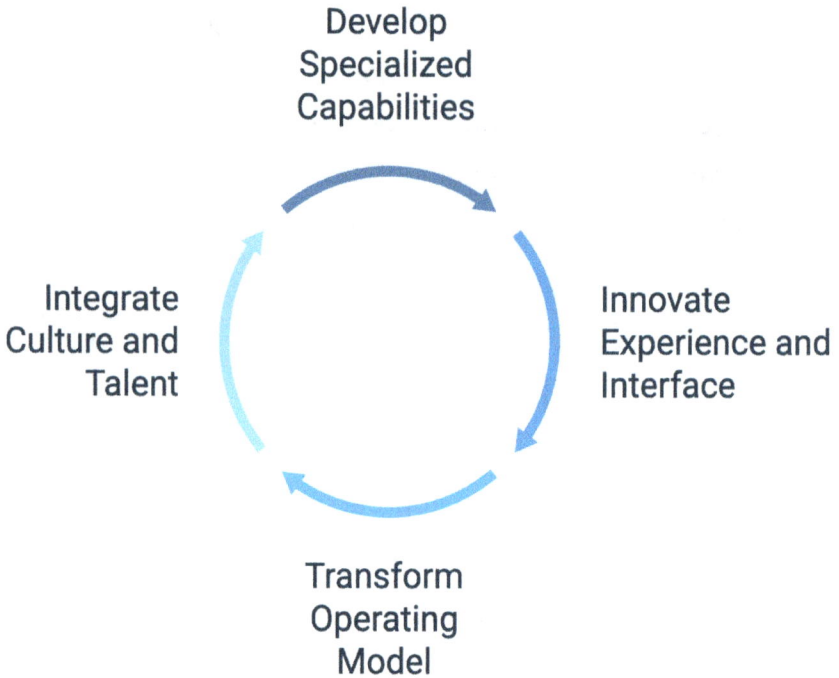

Develop
Specialized
Capabilities

Integrate
Culture and
Talent

Innovate
Experience and
Interface

Transform
Operating
Model

Strategic approach considerations

Organizations should develop explicit differentiation strategies based on their broader competitive positioning:

Alignment with existing advantage: Building digital workforce differentiation that enhances and extends current competitive strengths rather than pursuing unrelated capabilities.

Strategic intent reflection: Ensuring digital workforce approaches support strategic direction rather than creating disconnected technological excellence.

Customer value connection: Focusing differentiation on elements that create meaningful customer value rather than internal efficiency alone.

Sustainable advantage creation: Developing difficult-to-replicate capabilities through unique combinations of technology, process, data, and human factors.

Continuous evolution planning: Creating approaches that adapt as technology capabilities and competitive landscapes evolve rather than static advantage definitions.

Organizations that develop clear differentiation strategies typically capture greater value from digital workforce investments than those pursuing generic implementation approaches based primarily on vendor offerings.

Preparation Strategies: Building Adaptive Capacity

Organizations must develop the capacity to adapt to emerging capabilities, changing competitive dynamics, and shifting societal expectations.

Organizational Flexibility Mechanisms

To thrive amid rapid technological evolution, organizations must build structural and operational flexibility that enables responsive adaptation.

Flexibility enablement approaches

1. **Modular architecture development**

- Component-based system design
- Standardized interface definition
- Capability encapsulation
- Independent evolution enablement
- Reconfiguration flexibility

2. **Responsive resource allocation**

- Dynamic investment adjustment capability

- Rapid redeployment mechanisms
- Portfolio management flexibility
- Experimental funding approaches
- Value-based reprioritization systems

3. **Organizational structure adaptation**

- Network-based collaboration models
- Team formation and dissolution fluidity
- Role definition flexibility
- Authority distribution adaptability
- Cross-functional integration mechanisms

4. **Process evolution capability**

- Continuous improvement infrastructure
- Rapid experimentation approaches
- Learning integration mechanisms
- Performance feedback acceleration
- Adaptive policy frameworks

Pathways to Organizational Agility

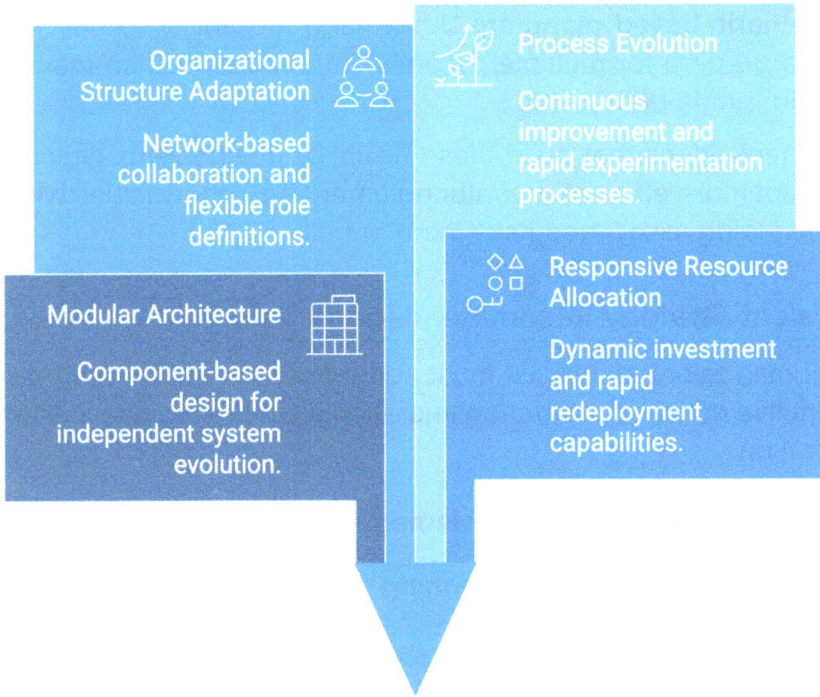

Organizational Structure Adaptation

Network-based collaboration and flexible role definitions.

Process Evolution

Continuous improvement and rapid experimentation processes.

Modular Architecture

Component-based design for independent system evolution.

Responsive Resource Allocation

Dynamic investment and rapid redeployment capabilities.

Implementation considerations

Several practical approaches can enhance organizational flexibility:

Minimum viable bureaucracy: Creating only essential governance and control structures while emphasizing principles over detailed rules where appropriate.

Capability-centered organization: Structuring around capabilities and value streams rather than rigid functional hierarchies to enable more fluid adaptation.

Responsible experimentation culture: Building norms and capabilities for rapid, low-risk experimentation and learning integration.

Distributed decision authority: Pushing appropriate decision rights to where information and expertise reside rather than requiring extensive approval chains.

Scenario-based planning: Developing the ability to envision and prepare for multiple potential futures rather than making rigid, single-path plans.

Organizations that build these flexibility mechanisms typically adapt more efficiently, capturing emerging opportunities while managing associated risks more effectively.

Talent Strategy Recommendations

Human talent strategies must adapt to focus on areas of distinctive human contribution and effective human-digital collaboration.

Talent strategy evolution elements

1. **Strategic workforce planning adaptation**

- Integrated human-digital capability planning
- Skill evolution anticipation
- Critical human capability identification
- Novel role definition approaches
- Transition pathway development

2. **Talent acquisition evolution**

- Complementary skill emphasis
- Adaptability and learning focus
- Human-digital collaboration aptitude
- Creative and judgment capability prioritization
- Technical-human bridging ability

3. Development approach transformation

- Continuous learning infrastructure
- Adaptation capability building
- Human advantage area enhancement
- Digital collaboration skill development
- Career transition support

4. Performance management reinvention

- Human-digital team effectiveness measurement
- Value contribution focus beyond task completion
- Adaptation and learning incentives
- Collaboration effectiveness evaluation
- Innovation and improvement recognition

Talent Strategy Evolution Cycle

Performance Management Reinvention

Measure team effectiveness and innovation

Strategic Workforce Planning

Anticipate skill needs and roles

Development Approach Transformation

Build continuous learning infrastructure

Talent Acquisition

Focus on adaptability and collaboration

Implementation priority areas

Organizations should prioritize several talent strategy elements:

Learning culture development: Building organizational norms and systems that enable continuous adaptation through systematic learning.

Human advantage investment: Focusing development resources on distinctly human capabilities unlikely to be replicated by digital workers.

Transition support creation: Developing programs and pathways for employees whose roles are significantly impacted by digital worker implementation.

Leadership capability evolution: Enhancing leadership development to build skills for guiding integrated human-digital teams.

Talent experience redesign: Creating employment experiences that attract and retain key talent in an environment of changing role definitions and skill requirements.

Organizations that proactively evolve talent strategies create powerful advantages through superior human capability development, more effective human-digital integration, and enhanced adaptability.

Investment Prioritization Frameworks

As digital workforce possibilities expand, organizations must make increasingly complex investment decisions across competing opportunities, requiring structured prioritization approaches.

Prioritization dimension framework

1. **Value creation potential**

- Direct financial return expectations
- Strategic positioning enhancement
- Capability-building contribution
- Risk management improvement

- Organizational learning value

2. **Implementation feasibility**

- Technical complexity assessment
- Organizational readiness evaluation
- Resource requirement appropriateness
- Risk and uncertainty level
- Integration complexity considerations

3. **Strategic alignment strength**

- Core business priority support
- Competitive advantage enhancement
- Customer value proposition advancement
- Long-term direction alignment
- Cultural value compatibility

4. **Option value generation**

- Future opportunity creation
- Flexibility enhancement
- Knowledge and capability building
- Strategic positioning improvement
- Ecosystem relationship development

Decision approach recommendations

Investment approach spectrum from rigid to flexible strategies

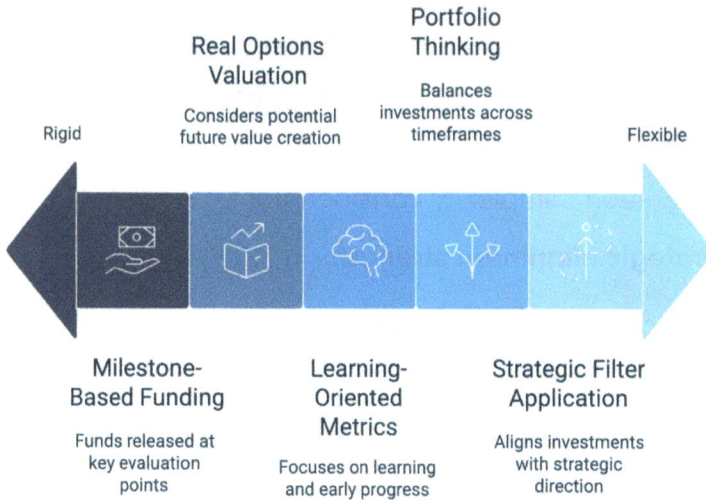

Real Options Valuation
Considers potential future value creation

Portfolio Thinking
Balances investments across timeframes

Rigid — Flexible

Milestone-Based Funding
Funds released at key evaluation points

Learning-Oriented Metrics
Focuses on learning and early progress

Strategic Filter Application
Aligns investments with strategic direction

Several decision-making approaches enhance investment prioritization effectiveness:

Portfolio thinking: Developing balanced digital workforce investment portfolios across different timeframes, risk levels, and strategic objectives rather than evaluating opportunities in isolation.

Real options valuation: Incorporating option value creation into investment assessment rather than relying solely on direct return calculations.

Learning-oriented metrics: Establishing explicit learning objectives and measures for early-stage investments rather than applying mature implementation performance expectations.

Milestone-based funding: Using stage-gate approaches with explicit evaluation points rather than full upfront commitment for unproven implementations.

Strategic filter application: Applying explicit strategic criteria to ensure investments support basic direction rather than pursuing technological implementation for its own sake.

Organizations that develop sophisticated prioritization frameworks typically achieve better returns on investments while building more coherent capabilities aligned with strategic direction.

Scenario Planning and Strategic Flexibility

Given the uncertainty in digital workforce evolution, organizations benefit from developing multiple potential future scenarios and creating strategies with appropriate flexibility to adapt as actual developments unfold.

Scenario development approaches

1. **Key uncertainty identification**

- Technology capability evolution possibilities
- Competitive landscape development options
- Regulatory environment scenarios
- Workforce impact and adaptation alternatives
- Customer and market acceptance variations

2. **Scenario construction methodologies**

- Alternative future narrative development
- Internal consistency verification
- Plausibility assessment
- Distinctive difference assurance
- Strategic implication exploration

3. **Robust strategy development**

- Common element identification across scenarios

- No-regret move recognition
- Flexibility value quantification
- Hedging approach development
- Option creation opportunity identification

Implementation recommendations

Organizations can enhance strategic flexibility through several practical approaches:

Leading indicator monitoring: Establishing metrics and monitoring systems to provide early signals about which scenarios are becoming more likely.

Strategic option development: Creating explicit options that can be exercised as the environment evolves rather than making irreversible commitments.

Minimum viable implementation: Designing initial deployments to provide maximum learning and adaptation opportunity while minimizing irreversible commitment.

Ecosystem positioning focus: Establishing positions in emerging ecosystems that maximize future flexibility while creating current capability access.

Capability-centered investment: Prioritizing foundational capability development that creates value across multiple potential futures rather than scenario-specific implementations.

Organizations that embrace uncertainty through scenario planning and strategic flexibility typically navigate emerging opportunities more effectively than those making rigid, single-future bets or delaying action until uncertainty resolves.

Conclusion:
Leading Through Technological Transformation

The evolution of digital workforces is one of the most significant technological transformations in decades, with profound implications for organizational strategy, operations, and competitive positioning.

As these technologies continue to advance, organizational leaders face complex challenges in navigating uncertain futures while capturing emerging opportunities.

Several principles can guide effective leadership through this transformation:

1. **Balance vision and pragmatism**: Combining an ambitious understanding of future possibilities with practical recognition of current limitations and implementation realities.

2. **Integrate technology and humanity**: Focusing on the unique combination of human and digital capabilities rather than viewing either in isolation or assuming simple substitution relationships.

3. **Develop adaptive foundations**: Building technical architecture, organizational capabilities, and governance approaches that can evolve as technologies and requirements change rather than optimizing exclusively for current conditions.

4. **Maintain ethical centrality**: Ensuring values and ethical considerations remain central to implementation decisions rather than treating them as secondary constraints or compliance requirements.

5. **Embrace continuous learning**: Establishing systematic approaches to learn from implementation experiences, evolving capabilities, and emerging market developments rather than executing static plans.

Organizations that navigate this transformation effectively will create significant advantages through enhanced productivity, improved customer experiences, new business model opportunities, and superior talent engagement. Those that fail to adapt risk competitive disadvantage and potential disruption.

The digital workforce revolution has only begun, with the most profound implications still emerging. Organizations that develop the capabilities to anticipate these developments, adapt as they unfold, and shape their direction will define the next era of business competition and organizational effectiveness.

CHAPTER 12:

YOUR DIGITAL WORKFORCE JOURNEY

TL;DR:

- Successful digital workforce journeys begin with a comprehensive readiness assessment across technical, cultural, and resource dimensions to establish realistic starting points and identify critical gaps.

- Effective roadmaps integrate short-term quick wins, medium-term capability building, and long-term transformation vision into a coherent progression that maintains momentum while building toward strategic objectives.

- Change leadership requires sophisticated approaches at executive, middle management, and frontline levels, combining compelling narratives with practical enablement to drive sustainable adoption.

- Comprehensive success measurement evolves from initial implementation metrics to sophisticated business impact, capability development, and competitive positioning assessment that guides ongoing investment and enhancement.

- Each organization's digital workforce journey will be unique; those that combine ambitious vision with pragmatic implementation, relentless value focus, and continuous learning will create significant and sustainable competitive advantage.

Building Your Digital Workforce Journey

The preceding chapters have explored the technological foundations, economic implications, implementation approaches, and management considerations for digital workforces.

This final chapter turns the focus directly to your organization's specific journey, how to assess readiness, develop a comprehensive roadmap, lead the necessary changes, and measure success in your digital workforce transformation.

Readiness Assessment: Evaluating Organizational Preparedness

A systematic assessment of the organization's current state of readiness across multiple dimensions is required before beginning implementation. This assessment provides crucial insights for effective planning and risk management.

Technical Readiness Evaluation Framework

Technical foundations significantly impact the success of digital workforce implementation. A comprehensive assessment examines infrastructure, data, and technical capabilities to identify strengths and gaps.

Key technical readiness dimensions

Agentic AI Technical Readiness

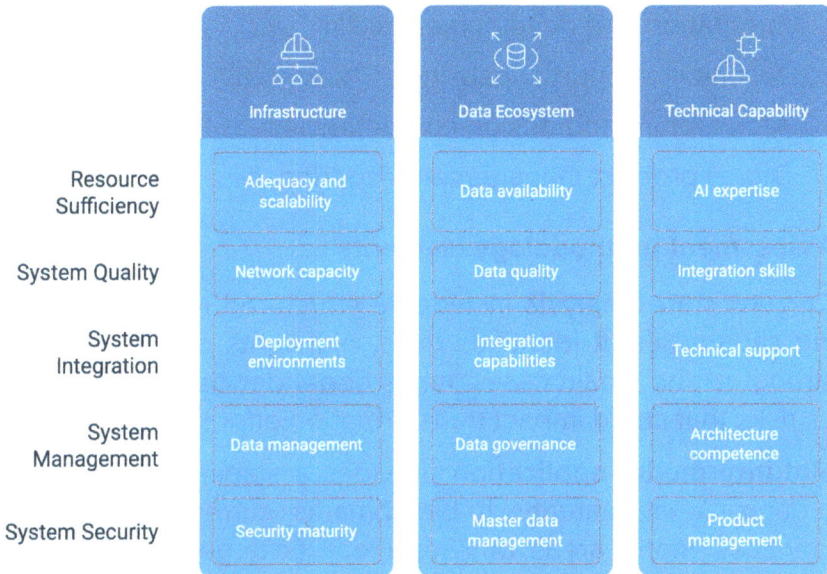

	Infrastructure	Data Ecosystem	Technical Capability
Resource Sufficiency	Adequacy and scalability	Data availability	AI expertise
System Quality	Network capacity	Data quality	Integration skills
System Integration	Deployment environments	Integration capabilities	Technical support
System Management	Data management	Data governance	Architecture competence
System Security	Security maturity	Master data management	Product management

1. Infrastructure preparedness

- Computing resource adequacy and scalability
- Network capacity and reliability
- Storage systems and data management capabilities
- Security infrastructure maturity
- Development and deployment environments

2. Data ecosystem maturity

- Data availability for key processes
- Data quality and consistency
- Integration capabilities across systems
- Master data management approaches
- Data governance frameworks

3. **Technical capability assessment**

- AI and automation expertise availability
- Integration and development skills
- Technical support capabilities
- Architecture and design competencies
- Digital product management experience

Assessment methodology elements

Current state analysis includes a structured review of existing capabilities. It identifies gaps, highlights patterns across strengths and weaknesses, maps dependencies, and assesses risks that could impact transformation efforts.

Maturity model application provides a stage-based view of organizational maturity. It benchmarks current capabilities, identifies priority areas, outlines a realistic target state, and defines a clear pathway for advancement.

Artificial Intelligence Organizational Maturity

Level 4 Strategic — Driving competitive differentiation and innovation: AI First

Level 3 Systemic — Process focused, integrated and embedded - goal aligned

Level 2 Operational — Productivity focused, lacking integration

Level 1 Opportunistic — Individual / departmental experimentation

Level 0 No Capabilities

Dimensions:
- Data
- Technology
- Governance
- People

Arion Research LLC

Artificial Intelligence Organizational Maturity

Level	Data	Technology	Governance	People
4	Complete, Integrated Data Infrastructure	AI enabled end-to-end business processes	Organizational governance, corporate ethics framework	CEO Exec Sponsor CAIO / CDO AI First Strategy
3	MLOps organizational data strategy, federated data	Systemic Embedded AI	Cross departmental governance	Interdepartmental, departmental exec sponsor(s)
2	Fragmented data strategy	Stand-a-lone Generative AI, Embedded ML / AI	Fragmented governance	Departmental
1	No organized data strategy	Consumer Generative AI	No governance	Individual / team

Arion Research LLC

Enhancement planning focuses on critical gap closure strategies. It involves prioritizing capability building, identifying resources required, developing timelines for foundation building, and creating risk mitigation approaches.

The most effective technical readiness assessments balance aspirational targets with practical constraints, creating realistic plans that acknowledge current limitations while establishing pathways for necessary evolution.

Cultural Readiness Assessment Methodology

Organizational culture dramatically impacts digital workforce adoption success. Systematic assessment of cultural dimensions enables more effective change management strategies.

Cultural readiness dimensions

1. **Change receptivity**

- Historical experience with transformation
- Adaptability to new ways of working
- Openness to technology evolution
- Continuous improvement orientation
- Learning culture presence

2. **Collaboration orientation**

- Cross-functional cooperation effectiveness
- Knowledge sharing practices
- Team-based working approaches
- Collaborative decision-making norms
- Boundary-spanning relationship strength

3. **Innovation culture**

- Experimentation comfort level
- Risk tolerance for new approaches
- Creative problem-solving encouragement
- Status quo questioning acceptance
- Failure learning integration

4. **Trust dynamics**

- Technology confidence levels
- Leadership credibility perception
- Change motivation transparency
- Psychological safety for expressing concerns
- Historical trust experience with technology initiatives

Cultural Readiness Dimensions

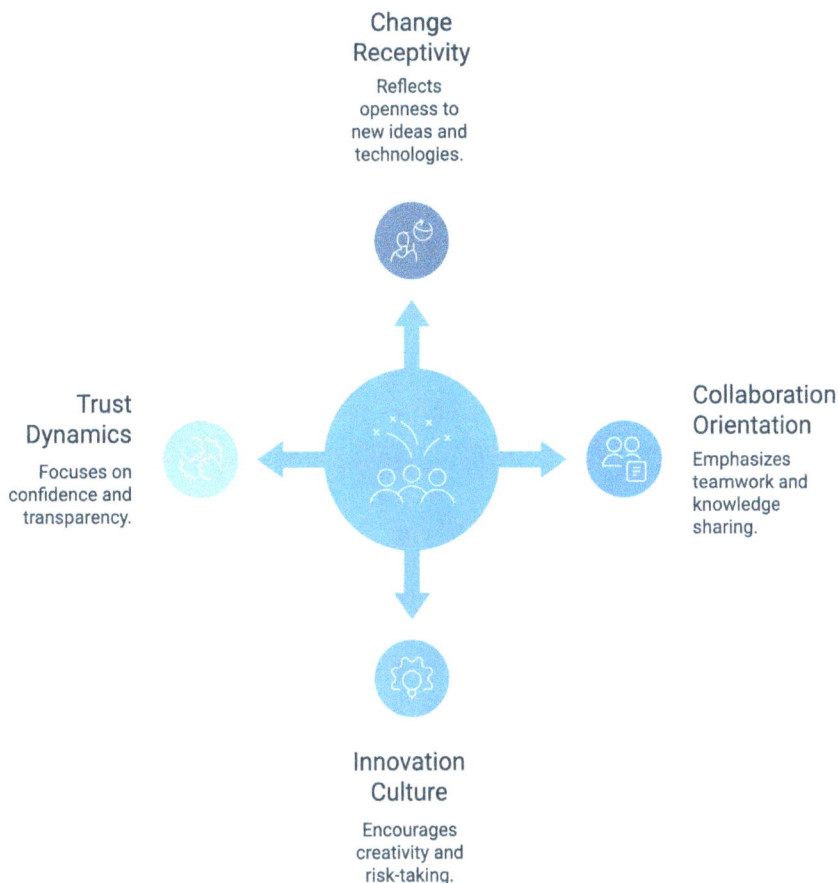

Change Receptivity
Reflects openness to new ideas and technologies.

Trust Dynamics
Focuses on confidence and transparency.

Collaboration Orientation
Emphasizes teamwork and knowledge sharing.

Innovation Culture
Encourages creativity and risk-taking.

Assessment approach elements

Multi-level analysis captures the perspectives of executive leadership, middle management, and front-line employees.

This approach helps reveal variations across functions and identifies distinct subcultures within the organization, offering a fuller picture of alignment and disconnects.

Mixed-method assessment combines diverse data sources for a well-rounded understanding. It uses quantitative survey instruments and qualitative focus group discussions.

Leadership interviews offer strategic insights, complemented by observational analysis of current behaviors and review of historical patterns to identify enduring dynamics.

Targeted insight development focuses on uncovering actionable findings. It involves identifying specific barriers to adoption, recognizing cultural enablers that support change, and deriving implications for change management.

These insights help refine communication approaches and shape tailored engagement strategies for more effective implementation.

Organizations that undergo mature cultural assessment typically develop more effective change management and implementation approaches by addressing specific cultural factors.

Resource Alignment Analysis

Digital workforce implementation requires appropriate alignment of financial, human, technological, and organizational resources. Effective assessment examines current allocation patterns and identifies necessary adjustments.

Resource dimension categories

1. **Financial resource alignment**

- Budget allocation patterns
- Investment priority alignment
- Funding flexibility and availability
- Cost management capabilities
- Value tracking mechanisms

2. **Human resource readiness**

- Skill availability and gaps
- Capacity for implementation support

- Role clarity and responsibility allocation
- Learning and development infrastructure
- Talent acquisition capabilities

3. **Organizational structure fitness**

- Decision-making process efficiency
- Accountability framework clarity
- Cross-functional coordination mechanisms
- Governance structure appropriateness
- Innovation support systems

4. **Partner ecosystem development**

- Strategic partner relationships
- Vendor management capabilities
- External expertise access
- Collaboration effectiveness
- Knowledge transfer mechanisms

Resource Dimension Categories

	Financial Resources	Human Resources	Organizational Structure	Partner Ecosystem
Budgeting	Allocation patterns	Skill availability	Decision-making efficiency	Strategic relationships
Investment	Priority alignment	Implementation support	Accountability clarity	Vendor management
Funding	Flexibility, availability	Role clarity	Coordination mechanisms	External expertise
Cost	Management capabilities	Learning infrastructure	Governance appropriateness	Collaboration effectiveness
Value	Tracking mechanisms	Talent acquisition	Innovation support	Knowledge transfer

Assessment methodology components

Current allocation analysis examines resource distribution patterns, assesses whether allocations align with strategic priorities, and identifies constraints or limitations. This process also reviews how efficiently resources are utilized and highlights gaps or opportunities for reallocation or optimization.

Future requirement projection involves anticipating what will be needed for successful implementation. It includes forecasting resource demands over time, identifying critical dependencies, assessing areas of potential risk exposure, and determining contingency plans to address uncertainties or disruptions.

Alignment planning focuses on ensuring resources match future needs. This includes creating reallocation strategies, planning acquisitions where gaps exist, determining optimal timing and sequencing, developing interim mitigation approaches, and defining a phased evolution pathway toward the target state.

Effective resource alignment helps organizations achieve more efficient implementation with fewer disruptions and resource conflicts than those that embark on digital workforce initiatives without this preparatory analysis.

Roadmap Development: Creating Your Implementation Strategy

Based on readiness assessment insights, organizations must develop comprehensive roadmaps that guide their digital workforce journey from initial exploration through mature implementation.

Identifying Short-Term Quick Wins

Early implementation successes build momentum, generate organizational learning, and create support for broader initiatives. Systematically identifying the right opportunities for these quick wins is a crucial planning element.

Quick win criteria framework

1. **Value potential**

- Meaningful business impact achievable
- Clear and measurable outcomes possible
- Direct connection to strategic priorities
- Visible benefits to key stakeholders
- Reasonable return on investment

2. **Implementation feasibility**

- Manageable technical complexity
- Limited dependency entanglement
- Available implementation capabilities
- Reasonable resource requirements
- Achievable within 3-6 months

3. **Organizational readiness**

- Sponsor/champion availability
- Stakeholder support presence
- Necessary subject matter expertise access
- Required system access and integration capability
- Change receptivity in affected areas

Quick Win Criteria Funnel

1 Value Potential
Assessing business impact and ROI

2 Implementation Feasibility
Evaluating technical and resource aspects

3 Organizational Readiness
Ensuring support and expertise

Identification methodology elements

Opportunity scanning approaches help uncover areas for improvement by analyzing process pain points, identifying user frustrations, recognizing resource bottlenecks, examining high-volume routine tasks, and spotting sources of value leakage.

Prioritization frameworks guide decision-making by calculating value-to-effort ratios, assessing strategic alignment, evaluating potential to build experience, considering associated risk levels, and minimizing dependencies.

Implementation planning involves defining the right team composition, allocating necessary resources, setting timelines and milestones, developing risk management strategies, and creating frameworks to measure success.

The most effective quick-win strategies balance meaningful value creation with execution confidence, avoiding both trivial implementations with limited impact and overly ambitious projects with high failure risk.

Medium-Term Capability Building Plan

Beyond initial quick wins, organizations must develop mid-term plans that systematically build capabilities and expand implementation scope in alignment with strategic priorities.

Capability development framework

1. **Building the technical foundation**

- Architecture and infrastructure evolution
- Data quality and availability enhancement
- Integration capability development
- Security and governance maturation
- Development environment improvement

2. **Organizational capability enhancement**

- Core team expansion and development
- Skill building across key functions
- Process redesign capability establishment
- Change management capacity building
- Implementation methodology refinement

3. **Skills development strategy**

- Pilot portfolio diversification
- Learning capture and dissemination
- Pattern recognition across implementations
- Best practice development
- Center of excellence establishment

Agentic AI Capability Development Framework

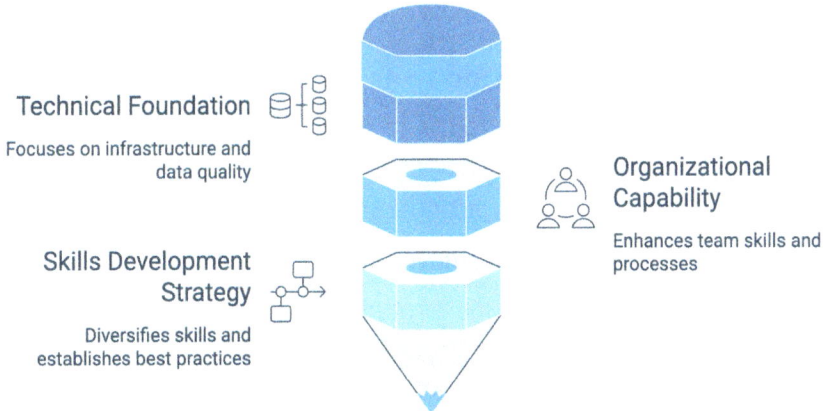

Technical Foundation
Focuses on infrastructure and data quality

Organizational Capability
Enhances team skills and processes

Skills Development Strategy
Diversifies skills and establishes best practices

Implementation expansion approach

Scope progression planning involves sequencing function-wise expansion, establishing a complexity gradient, ordering implementation based on dependency, designing a value maximization pathway, and balancing risks across the portfolio.

Scale development strategy involves setting up a pilot to production methodology, standardizing for efficiency, and creating reusable components. It also includes building mechanisms for knowledge transfer and expanding support capabilities.

Integration depth evolution starts with focusing on assistant or augmentation capabilities, then progressively expanding autonomy. It refines human-digital workflows, builds performance optimization approaches, and develops experience personalization elements.

A well-structured medium-term plan provides clear direction while maintaining appropriate flexibility to adapt as implementation experience accumulates and organizational knowledge expands.

Long-Term Transformation Vision

A compelling long-term vision articulates how digital workers will transform operations, offerings, and competitive positioning.

Vision development components

1. **Future state definition**

- Digital workforce role articulation
- Human-digital collaboration model
- Organizational structure evolution
- Process transformation description
- Culture and way-of-working changes

2. **Value creation framework**

- Customer experience enhancement vision
- Operational excellence transformation
- Business model evolution opportunities
- Employee experience reimagination

- Competitive differentiation approach

3. **Strategic positioning evolution**

- Industry leadership aspiration
- Capability-centered advantage definition
- Partnership ecosystem development
- Talent strategy transformation
- Innovation acceleration approach

Achieving Strategic Vision

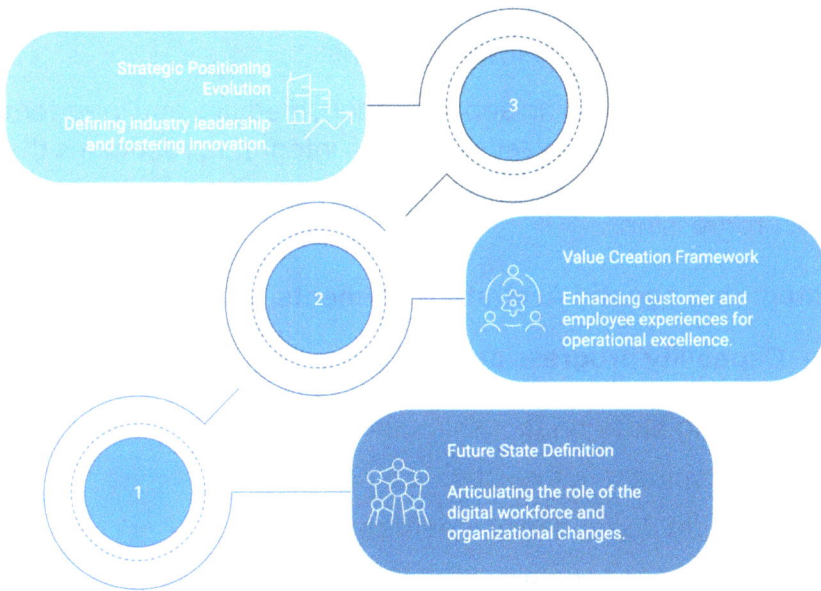

Strategic Positioning Evolution
Defining industry leadership and fostering innovation.

3

2

Value Creation Framework
Enhancing customer and employee experiences for operational excellence.

1

Future State Definition
Articulating the role of the digital workforce and organizational changes.

Vision application elements

Communication and alignment involve developing a compelling narrative, building stakeholder-specific messaging, creating concrete examples, creating regular reinforcement mechanisms, and connecting progress to the destination.

Decision guidance function involves a framework for investment prioritization and a mechanism for trade-off resolutions. It includes guidance for resource allocation, criteria for evaluating initiatives, and defines course correction triggers.

Adaptation mechanisms involve regular review and refinement processes, learning integration approaches, and responses to environmental changes. It also incorporates emerging opportunities and focuses on continuous evolution rather than a static endpoint.

The most effective transformation visions combine ambitious aspiration with practical credibility, inspiring organizational commitment while providing meaningful guidance for implementation decisions.

Integration Across Time Horizons

Rather than developing separate short, medium, and long-term plans, effective organizations create integrated roadmaps that connect immediate actions to long-term vision through coherent progression.

Integrated agentic AI strategy elements

1. **Capability progression linkage**

- Building block relationship definition
- Foundation to future-state connection
- Learning dependency identification
- Prerequisite capability mapping
- Maturity evolution visualization

2. **Value realization continuity**

- Progressive benefit accumulation planning
- Investment return timeframe balancing
- Quick win to strategic advantage connection
- Stakeholder value delivery sequencing

- Sustainable momentum maintenance

3. **Risk management across horizons**

- Early de-risking of critical components
- Experimental approach to uncertain elements
- Option-creating implementation design
- Path flexibility preservation
- Reversibility consideration for major commitments

Integrated AI Strategy Cycle

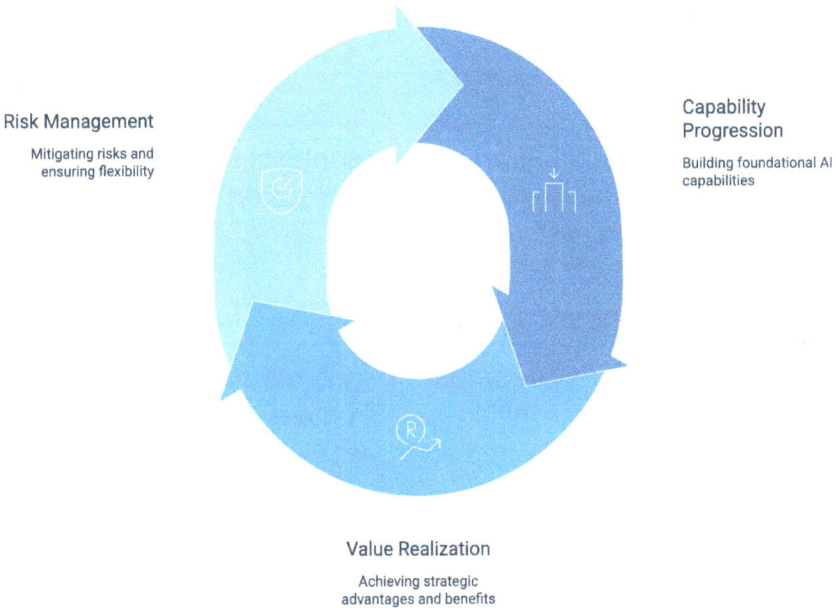

Risk Management

Mitigating risks and
ensuring flexibility

Capability
Progression

Building foundational AI
capabilities

Value Realization

Achieving strategic
advantages and benefits

Implementation approach considerations

Rolling planning methods establish a cadence for regularly refreshing plans, combining near-term detail with longer-term direction. They include mechanisms for incorporating learning, making adjustments without abandoning the overall vision, and responding to opportunities while maintaining strategic focus

Milestone-based governance focuses on identifying decision points, pre-defining evaluation criteria, establishing go/no-go gates, creating opportunities for path adjustment, and outlining achievement celebration points.

Stakeholder journey management involves designing experiences across the timeline, setting and managing expectations, ensuring progress visibility, maintaining consistent engagement, and building approaches for continuous communication.

Organizations that develop these integrated roadmaps typically maintain better implementation momentum, resource alignment, and stakeholder support than those with disconnected planning horizons or excessive rigidity.

Change Leadership: Driving Successful Transformation

Digital workforce initiatives require sophisticated change leadership to overcome resistance, build support, and create sustainable adoption.

Executive Sponsorship Strategies

Executive support is perhaps the most critical success factor for digital workforce initiatives. Effective implementation requires specific sponsorship approaches beyond generic leadership endorsement.

Sponsorship role framework

1. **Strategic direction setting**

- Vision articulation and reinforcement
- Priority and focus area definition
- Resource commitment authorization
- Cross-functional alignment creation
- Ambition level calibration

2. **Barrier removal and protection**

- Organizational obstacle elimination
- Resource conflict resolution
- Political resistance management
- Organizational noise filtering
- Implementation space creation and preservation

3. **Cultural tone**

- Personal adoption demonstration
- Learning orientation modeling
- Appropriate risk-taking encouragement
- Failure response pattern setting
- Experimentation norm creation

4. **Progress accountability maintenance**

- Regular review cadence establishment
- Commitment fulfillment verification
- Course correction initiation
- Success recognition and reinforcement
- Continuous improvement expectation

Sponsorship development approaches

Executive engagement methods focus on connecting initiatives to personal relevance, explaining strategic linkages, demonstrating competitive implications, sharing success stories, and creating hands-on experiences.

Support capability building includes developing the key message, preparing for objections, providing data and evidence, creating talking points, and conducting regular briefings and updates.

Visible action orchestration involves identifying symbolic behaviors, leveraging high-visibility decisions, enabling public commitments, demonstrating personal investment, and reinforcing consistent messaging.

Organizations with strong executive sponsorship approaches typically achieve higher implementation success rates than those considering sponsorship a one-time endorsement rather than ongoing engagement.

Middle Management Enablement Approaches

While executive sponsorship provides crucial direction and support, middle management ultimately determines implementation success through day-to-day decisions, resource allocation, and team direction.

Middle management role framework

1. **Implementation translation**

- Strategic direction to operational action connection
- Team-specific implication explanation
- Concrete application identification
- Individual role impact clarification
- Daily work connection creation

2. **Resource orchestration**

- Practical allocation decisions
- Priority balancing management
- Time and attention direction
- Skill application facilitation
- Implementation support provision

3. **Resistance management**

- Concern identification and addressing
- Fear and uncertainty reduction
- Question response and clarification
- Misinformation correction
- Change rationale reinforcement

4. **Progress facilitation**

- Barrier identification and removal
- Implementation problem solving
- Cross-team coordination
- Learning and adjustment encouragement
- Momentum maintenance

Enablement strategy elements

Understanding and conviction building require early involvement in planning. It involves creating transparent business cases, soliciting questions and concerns, providing detailed briefings, and sharing peer experiences.

Capability development focuses on building digital workforce knowledge, developing change leadership and implementation management skills, strengthening team coaching skills, and providing resistance management techniques.

Incentive and motivation alignment includes adjusting performance expectations, recognizing support and progress, linking efforts to career opportunities, ensuring resources for implementation, and managing competing priorities.

Organizations that invest in comprehensive middle management enablement typically experience smoother implementation with less localized resistance, more consistent progress, and better operational integration.

Driving Frontline Adoption

Ultimately, digital workforce value depends on effective adoption and utilization by frontline employees whose daily work changes through implementation.

Adoption challenge categories

1. **Knowledge and skill gaps**

- System operation understanding
- New process familiarity
- Digital collaboration capability
- Problem-solving approach shifts
- Effective utilization knowledge

2. **Motivational barriers**

- Personal impact concerns
- Change fatigue effects
- Comfort with current approaches
- Risk and uncertainty perception
- Effort versus benefit assessment

3. **Practical implementation obstacles**

- Time for learning and adaptation
- Transition period complexity
- Early performance dips
- Integration with other systems and processes
- Technical issues and limitations

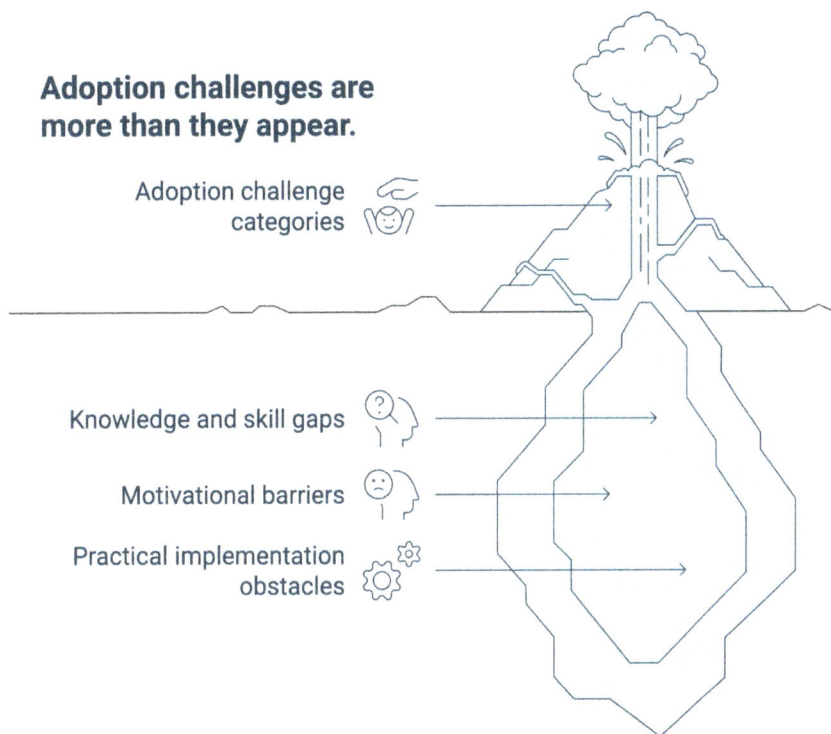

Adoption challenges are more than they appear.

Adoption challenge categories

Knowledge and skill gaps

Motivational barriers

Practical implementation obstacles

Facilitation strategy elements

Comprehensive training approaches focus on designing role-specific learning, emphasizing hands-on practice, providing just-in-time knowledge, offering self-service resources, and supporting continuous skill growth.

User experience optimization enhances interface usability, improves workflow integration, introduces complexity gradually, builds error-handling mechanisms, and incorporates user feedback for enhancement.

Peer support development includes creating champion and super-user networks, establishing communities of practice, encouraging knowledge sharing and peer-to-peer assistance, and spreading success stories.

Benefit demonstration and reinforcement ensure users see personal value, celebrate early wins, quantify time savings, highlight frustration reduction, and connect adoption to career growth.

Organizations with a focused adoption program typically achieve higher utilization levels, more consistent application, and greater value capture than those focusing exclusively on technical implementation without equivalent attention to human adoption factors.

Organizational Narrative Development

Successful digital workforce implementation requires compelling narratives that explain the why, what, and how of change in ways that build understanding and commitment across stakeholder groups.

Agentic AI narrative component framework

1. **Change rationale communication**

- External pressure explanation
- Opportunity articulation
- Risk avoidance description
- Strategic necessity connection
- Competitive implication explanation

2. **Future state visualization**

- Daily experience description
- Organizational capability illustration
- Customer impact explanation
- Competitive position portrayal
- Career opportunity presentation

3. **Journey description**

- Major milestone identification

- Timeline and progression explanation
- Learning approach articulation
- Support and assistance description
- Involvement opportunity explanation

Agentic AI Narrative Framework

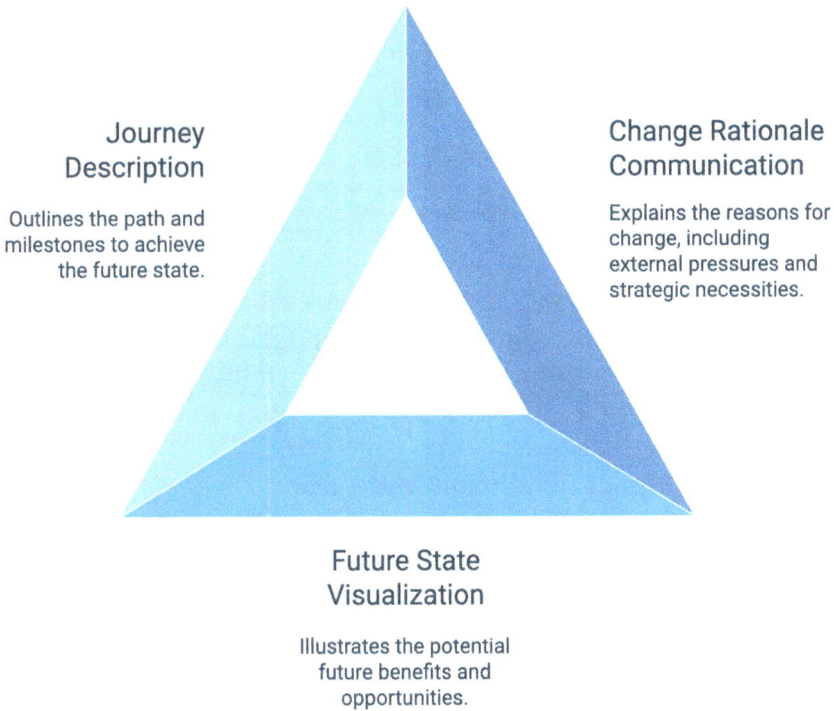

Journey Description

Outlines the path and milestones to achieve the future state.

Change Rationale Communication

Explains the reasons for change, including external pressures and strategic necessities.

Future State Visualization

Illustrates the potential future benefits and opportunities.

Narrative development approaches

Audience-specific tailoring ensures relevance by adapting messages to each stakeholder, calibrating detail, customizing examples, and choosing optimal communication mediums and channels.

Multi-voice orchestration aligns leadership messaging, adds functional depth, brings in peer views, includes third-party validation, and reflects customer and partner perspectives.

Experience-based enhancement strengthens engagement through pilot stories, user testimonials, real-world examples, solution-oriented narratives, and milestone celebrations.

The most effective organizational narratives combine rational business case elements with emotional engagement, creating both intellectual understanding and personal commitment to the digital workforce journey.

Measuring Success: Defining What Winning Looks Like

Effective digital workforce implementation requires clear definitions of success and systematic approaches to measuring progress and impact across multiple dimensions.

Business Impact Metrics

The ultimate measure of digital workforce success lies in tangible business impact. Comprehensive measurement approaches examine multiple impact dimensions rather than focusing on isolated metrics.

Agentic AI impact dimension framework

1. **Operational performance enhancement**

- Efficiency and productivity improvement
- Quality and accuracy advancement
- Throughput and capacity increase
- Response time reduction
- Consistency and reliability enhancement

2. **Financial impact measurement**

- Cost reduction realization
- Revenue enhancement contribution
- Capital efficiency improvement

- Risk exposure reduction
- Return on investment calculation

3. **Customer experience improvement**

- Satisfaction and loyalty enhancement
- Issue resolution acceleration
- Personalization capability expansion
- Availability and accessibility increase
- Problem prevention improvement

4. **Employee experience enhancement**

- Engagement and satisfaction impact
- Role evolution and enrichment
- Productivity tool provision
- Frustration and burden reduction
- Development opportunity creation

Exploring the Multifaceted Impact of Agentic AI

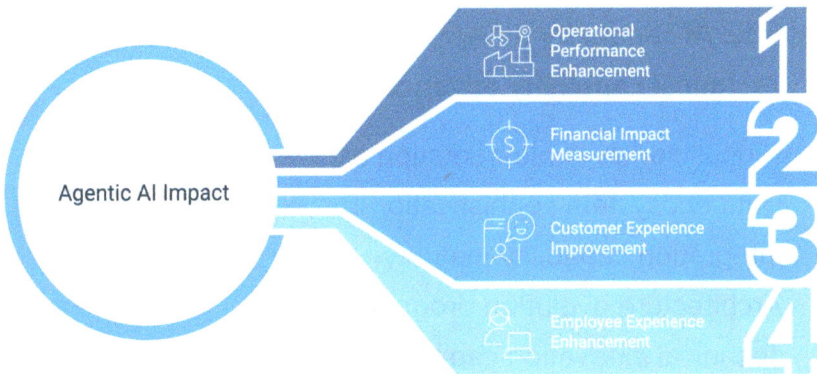

Agentic AI Impact

1 Operational Performance Enhancement

2 Financial Impact Measurement

3 Customer Experience Improvement

4 Employee Experience Enhancement

Measurement approach elements

Baseline establishment involves documenting pre-implementation performance, identifying comparison groups where possible, recognizing seasonal and cyclical patterns, standardizing current state metrics, and creating mechanisms for consistent measurement.

Attribution methodology focuses on applying isolation techniques, identifying control factors, normalizing external influence, linking outcomes to implementation timing, and examining alternative explanations.

Comprehensive valuation involves quantifying tangible benefits and recognizing intangible impacts. It calculates time-adjusted values, considers risk avoidance, and incorporates the strategic value of future options for a holistic view.

Organizations that have mature capabilities for measuring business impact typically make better ongoing investment decisions, create stronger implementation support, and identify enhancement opportunities more effectively.

Organizational Capability Development Indicators

Beyond immediate business impact, digital workforce initiatives should build lasting organizational capabilities that enable ongoing evolution and value creation.

Agentic AI capability development dimension framework

1. **Technical capability advancement**

- Digital worker implementation expertise
- Data ecosystem sophistication
- Integration capability maturity
- Architecture evolution progress
- Technical innovation capacity

2. **Process transformation progress**

- Redesign methodology sophistication
- Human-digital workflow optimization

- Cross-functional process integration
- End-to-end process ownership
- Continuous improvement mechanism effectiveness

3. **People capability enhancement**

- Digital literacy advancement
- Human-digital collaboration skill development
- Leadership capability evolution
- Innovation and creativity expansion
- Learning and adaptation acceleration

4. **Governance maturity progression**

- Decision-making efficiency improvement
- Risk management sophistication
- Resource allocation effectiveness
- Cross-functional coordination
- Responsible innovation capability

Achieving Agentic AI Capability

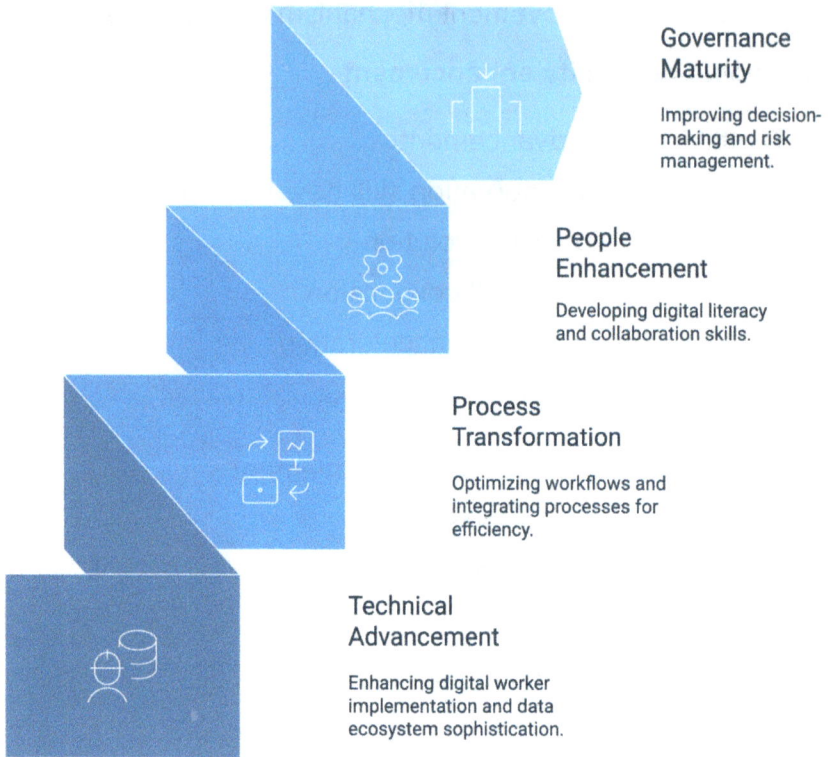

Governance Maturity

Improving decision-making and risk management.

People Enhancement

Developing digital literacy and collaboration skills.

Process Transformation

Optimizing workflows and integrating processes for efficiency.

Technical Advancement

Enhancing digital worker implementation and data ecosystem sophistication.

Measurement methodology elements

Maturity model application includes stage-based progression tracking, capability evolution mapping, benchmark comparisons, practice gap analysis, and development pathway visualization.

Learning effectiveness assessment evaluates knowledge capture, measures cross-initiative transfer, reduces repetitive mistakes, and accelerates the application of innovations. It also focuses on improving implementation efficiency.

Organizational flexibility measurement includes assessing adaptation speed, tracking how fast experiments scale, and evaluating response to novel challenges. It recognizes and captures opportunities as they arise, and gauges strategic evolution capacity.

Organizations that systematically track capability development typically sustain digital workforce value creation longer, adapt more effectively to changing conditions, and capture emerging opportunities more efficiently than those focusing exclusively on immediate business impact.

Competitive Positioning Assessment

Digital workforce initiatives should enhance organizational competitive position through unique combinations of capability, experience, and offering enhancements.

Positioning dimension framework

1. **Customer value proposition enhancement**

- Experience differentiation
- Quality and reliability advancement
- Price-performance improvement
- Novel capability introduction
- Relationship enhancement

2. **Cost position transformation**

- Structural cost advantage creation
- Scale economy enhancement
- Resource utilization optimization
- Value chain integration improvement
- Fixed-variable cost restructuring

3. **Agility and adaptation advantage**

- Market response acceleration
- Personalization at scale capability

- Innovation cycle compression
- Resource redeployment flexibility
- Learning and evolution speed

4. **Ecosystem position strengthening**

- Partner integration enhancement
- Data advantage accumulation
- Network effect acceleration
- Complementary offering expansion
- Standard and platform influence

Strategic Positioning Framework

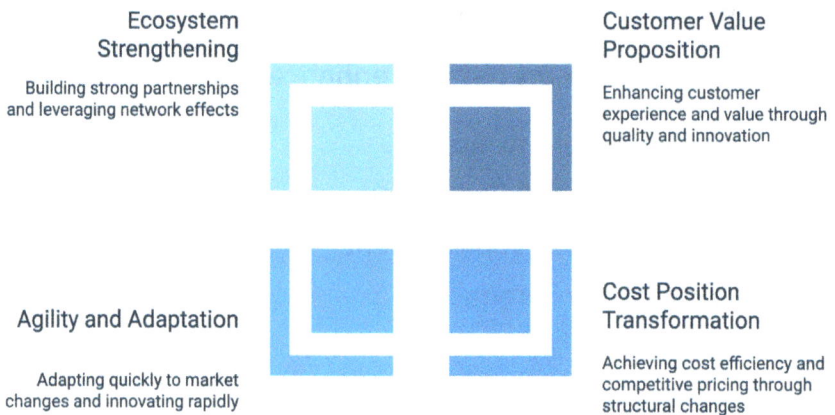

Ecosystem Strengthening

Building strong partnerships and leveraging network effects

Customer Value Proposition

Enhancing customer experience and value through quality and innovation

Agility and Adaptation

Adapting quickly to market changes and innovating rapidly

Cost Position Transformation

Achieving cost efficiency and competitive pricing through structural changes

Assessment approach elements

1. **Comparative analysis methods**

- Competitor capability benchmarking
- Customer perception measurement
- Analyst and market assessment
- Performance metric comparison
- Strategic position evaluation

2. **Evolution tracking approaches**

- Position trajectory mapping
- Capability gap closure measurement
- Differentiation sustainability assessment
- Advantage extension monitoring
- New position creation evaluation

3. **Market validation mechanisms**

- Customer choice analysis
- Win-loss pattern examination
- Price realization assessment
- Growth and share measurement
- Loyalty and retention analysis

Organizations with sophisticated competitive positioning assessment typically make more strategically coherent digital workforce investments, maintain clearer differentiation focus, and create more sustainable advantages than those with more internally focused measurement approaches.

Success Measurement Evolution

As digital workforce implementations mature, success measurement approaches should evolve from initial adoption metrics to sophisticated business impact and capability development assessment.

Measurement evolution framework

1. **Exploratory phase metrics**

- Pilot completion and functionality
- User adoption and engagement
- Basic performance verification

- Problem identification and resolution
- Learning capture and application

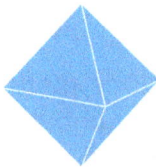

2. **Expansion phase measurement**

- Implementation scale and scope
- Performance consistency across contexts
- Efficiency and quality improvement
- Integration effectiveness
- User satisfaction and feedback

3. **Transformation phase assessment**

- Business process reinvention
- Customer experience transformation
- Business model evolution
- Competitive position enhancement
- Cultural and organizational change

Measurement Evolution Framework

Exploratory Phase	Expansion Phase	Transformation Phase	Enhanced Business
Initial metrics for pilot projects	Scaling and improving performance	Reinventing business and customer experience	Improved processes and customer experience

Implementation considerations

Measurement system evolution includes defining and refining metrics, expanding data collection capabilities, advancing analytical sophistication, enhancing insight development, and improving decision support systems.

Insight integration approaches involve using learning feedback loops, building strategy refinement mechanisms, adjusting resource allocation, evolving implementation methods, and identifying future opportunities.

Governance integration focuses on developing executive dashboards, establishing regular review cadences, linking decisions to measurements, reinforcing accountability, and enabling transparent reporting mechanisms.

Organizations with evolving measurement systems typically sustain implementation momentum longer, make better course corrections, and ultimately deliver greater value than those with static measurement approaches that fail to adapt as implementations mature.

Conclusion: Your Unique Journey

The digital workforce transformation is a profound shift in how organizations operate, create value, and compete. While this book has provided frameworks, methodologies, and guidance applicable across contexts, each organization's journey will be unique, shaped by its strategy, culture, capabilities, and market context.

Several principles can guide this unique journey:

1. **Start where you are**: Begin with an honest assessment of current readiness rather than aspirational self-perception, creating realistic implementation plans that acknowledge starting point realities.

2. **Focus on value creation**: Maintain relentless emphasis on business and customer value rather than technology

implementation for its own sake, ensuring digital workers enhance rather than complicate core operations.

3. **Balance ambition and pragmatism**: Combine ambitious vision of future possibilities with pragmatic recognition of current limitations, creating roadmaps that stretch the organization without breaking it.

4. **Integrate technology and humanity**: Recognize that successful implementation requires equal attention to technological and human dimensions, creating systems that enhance rather than diminish human capabilities.

5. **Learn and adapt continuously**: Approach the journey with humility and learning orientation, systematically capturing insights and evolving approaches rather than rigidly following initial plans.

Every organization will face unique challenges and discover distinctive opportunities along its digital workforce journey. Those that approach this transformation thoughtfully, with appropriate assessment, well-designed roadmaps, effective change leadership, and comprehensive success measurement, will create significant and sustainable value.

The organizations that thrive in the emerging era will be those that view digital workers not merely as cost-reduction tools but as essential capabilities that enable new levels of performance, customer experience, and competitive advantage. We hope this book serves as a valuable companion on your unique journey toward building a truly exceptional digital workforce.

CONCLUSION: LEADING IN THE AGE OF THE DIGITAL WORKFORCE

TL;DR:

- Successful digital workforce implementation follows five essential principles: focusing on business value over technology, integrating digital workers into organizational systems, creating human-digital synergy, emphasizing adaptability and learning, and implementing responsibly.

- Leaders guiding this transformation must develop critical capabilities, including strategic foresight amid uncertainty, cross-domain integration, balanced trust and oversight, learning leadership, and relentless focus on human development.

- Organizations should prepare for an accelerating future with expanding AI capabilities, evolving ecosystem dynamics, accelerating workforce transformation, shifting bases of strategic differentiation, and growing ethical and societal considerations.

- Long-term success depends on viewing implementation as a marathon rather than a sprint, creating value that builds organizational permission, maintaining people at the center, prioritizing integration over isolated implementation, and using organizational purpose to guide direction.

- The organizations that thrive will be those that thoughtfully balance technological possibility with human centrality, short-term progress with long-term vision, and efficiency with broader value creation.

The Hybrid Workforce

Throughout this book, we have explored the multifaceted dimensions of building, implementing, and managing digital workforces powered by agentic AI.

The preceding chapters have provided the frameworks and methodologies to enable this profound transformation, from understanding the technological foundations to navigating strategic, operational, and organizational implications.

As we conclude, it is worth reflecting on the key insights, the leadership imperatives, and the broader implications for the future of work and organizations.

Recap of Key Principles: Essential Insights for Digital Workforce Success

Several key principles have emerged across various domains that transcend specific implementation contexts and provide enduring guidance for digital workforce initiatives.

Technology as Enabler, Not Driver

Perhaps the most critical insight is that successful digital workforce implementation begins with **business value and human needs** rather than technology capabilities. Organizations that thrive in this transformation maintain relentless focus on where and how digital workers can create meaningful value.

This principle manifests in several important practices:

- Starting with clear business objectives rather than technological possibilities
- Designing digital workers to enhance rather than simply replace human capabilities

- Maintaining focus on customer and employee experience rather than internal efficiency alone
- Evolving implementation based on value realization rather than technical sophistication
- Measuring success through business impact rather than implementation scale

The organizations that maintain this value-centered approach typically achieve more sustainable impact, greater stakeholder support, and better return on investment than those driven primarily by technological fascination.

Integration Over Isolation

Successful digital workforces function as integral components of organizational systems rather than isolated technological implementations. This integration spans multiple dimensions.

- **Workflow integration**: Connecting digital and human activities into coherent processes rather than creating parallel operations
- **System integration**: Establishing effective connections with existing enterprise applications, data sources, and technical infrastructure
- **Organizational integration**: Embedding digital workers into organizational structures, reporting relationships, and governance frameworks
- **Strategic integration**: Aligning digital workforce initiatives with broader organizational priorities and direction
- **Cultural integration**: Evolving values, behaviors, and norms to incorporate digital workforce realities

Pathways to Digital Workforce Success

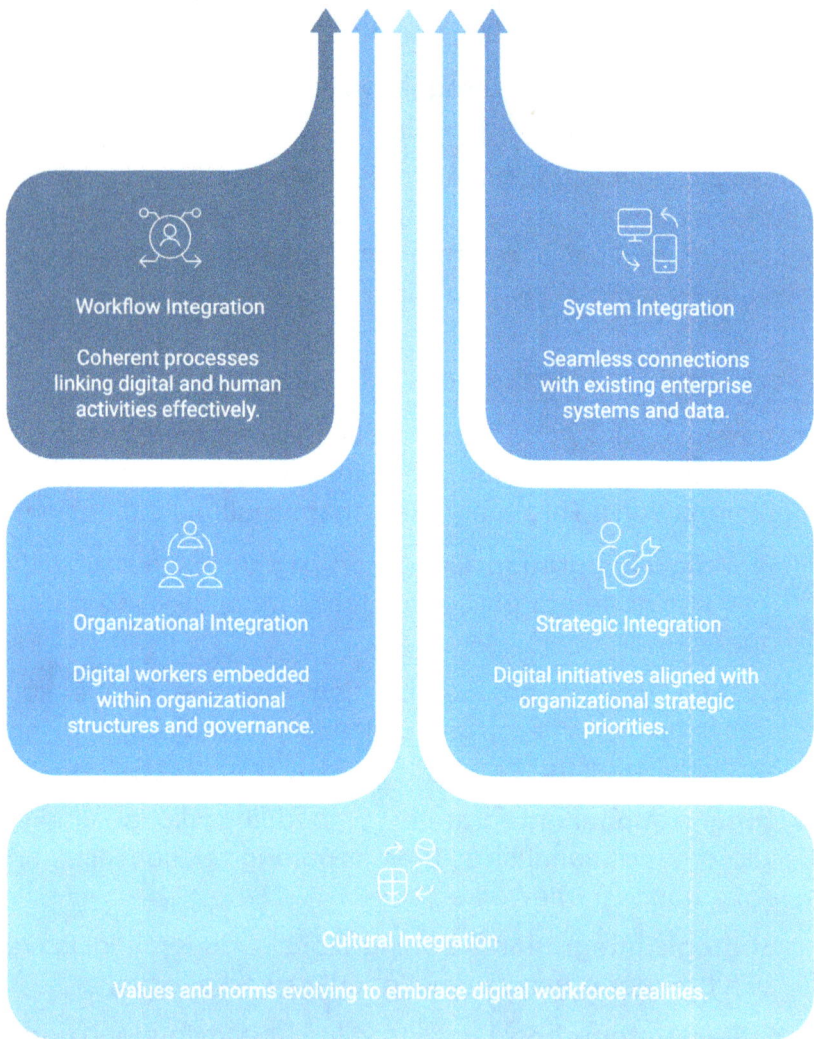

Workflow Integration

Coherent processes linking digital and human activities effectively.

System Integration

Seamless connections with existing enterprise systems and data.

Organizational Integration

Digital workers embedded within organizational structures and governance.

Strategic Integration

Digital initiatives aligned with organizational strategic priorities.

Cultural Integration

Values and norms evolving to embrace digital workforce realities.

Organizations that approach digital workers as integrated elements of their broader socio-technical systems typically achieve greater value, adoption, and sustainability than those implementing them as standalone capabilities disconnected from organizational context.

Human-Digital Synergy

The greatest value emerges from a thoughtful combination of human and digital capabilities that leverages the unique strengths of each. This synergistic approach requires:

- Recognizing the complementary nature of human creativity, judgment, and empathy alongside digital consistency, scalability, and analytical capability

- Designing collaborative workflows that enable seamless interaction and handoff between human and digital contributors

- Evolving human roles to emphasize distinctly human capabilities rather than competing with digital strengths

- Creating cultural norms and practices that support effective human-digital teaming

- Developing metrics and management approaches that encourage collaboration rather than competition

Organizations that master this synergistic approach create competitive advantages that are difficult to replicate, as they combine technological capability with uniquely human strengths in ways that pure automation or traditional human workforces cannot match.

Adaptability and Learning

Given the rapidly evolving nature of AI capabilities, regulatory landscapes, and competitive dynamics, successful digital workforce initiatives prioritize adaptability and continuous learning over rigid planning and static implementations.

This adaptive approach includes:

- Creating flexible technical architectures that incorporate emerging capabilities

- Establishing systematic learning processes that capture implementation insights

- Developing governance approaches that evolve as capabilities and requirements change
- Building teams with learning orientation and adaptation skills
- Implementing in ways that create options rather than constraints for future evolution

Foundations of Adaptable Digital Workforce

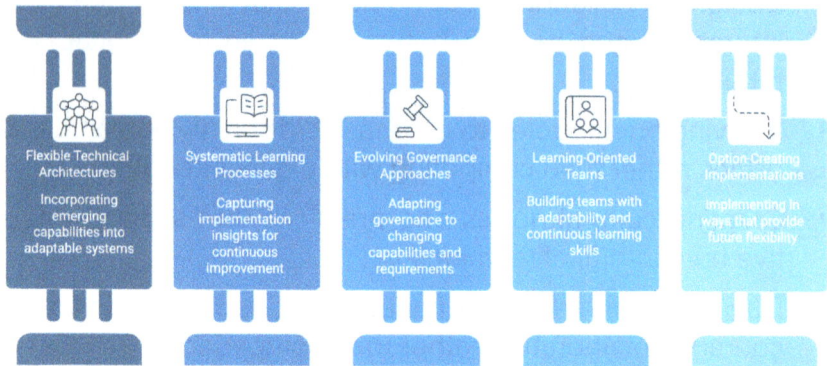

Organizations that embed this adaptability principle typically navigate technological evolution more effectively, respond to changing requirements more efficiently, and sustain competitive advantage longer than those with more rigid implementation approaches.

Responsible Implementation

Finally, successful digital workforce initiatives recognize the profound ethical, social, and organizational implications of these technologies and implement them in ways that align with broader organizational and societal values.

This responsible approach encompasses:

- Thoughtful consideration of workforce impacts and transition management

- Transparent communication with stakeholders about capabilities, limitations, and impacts
- Ethical frameworks that guide implementation decisions and boundary setting
- Governance mechanisms that ensure appropriate oversight and accountability
- Ongoing assessment of broader societal and economic implications

Responsible Digital Workforce Implementation

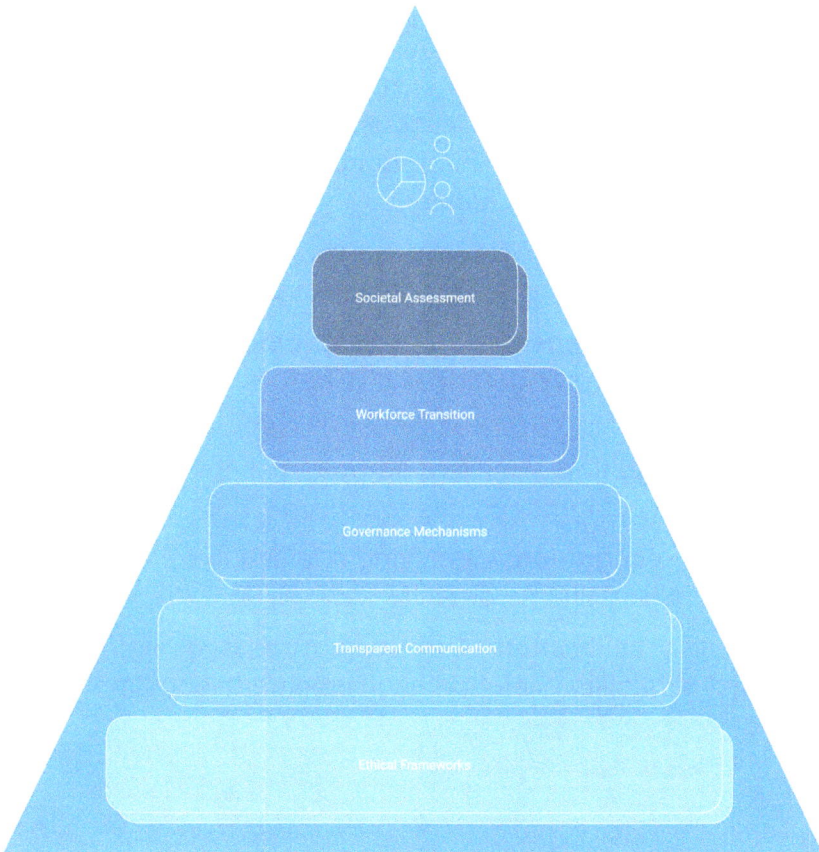

Organizations that embrace responsible implementation typically build greater trust with stakeholders, reduce implementation risks, and create more sustainable outcomes than those focusing exclusively on immediate business benefits.

Leadership Imperatives: Critical Capabilities for Executives

The digital workforce era demands new leadership capabilities that extend beyond traditional technology implementation or change management skills.

Strategic Foresight in Uncertainty

Leaders must navigate profound uncertainty about technology evolution, competitive implications, regulatory developments, and workforce impacts. This requires sophisticated strategic foresight capabilities, including:

- **Scenario thinking**: Developing multiple future scenarios rather than single-point predictions to guide planning
- **Opportunity sensing**: Identifying emerging possibilities created by evolving capabilities before they become obvious
- **Risk recognition**: Anticipating potential challenges, limitations, and unintended consequences
- **Pattern recognition**: Identifying relevant insights from early signals and limited data
- **Timeline calibration**: Developing realistic expectations about capability evolution and implementation timeframes

Strategic Foresight Process

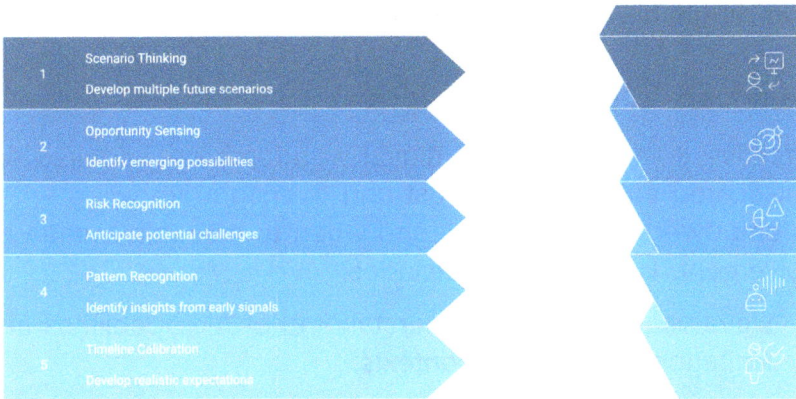

1	**Scenario Thinking** — Develop multiple future scenarios
2	**Opportunity Sensing** — Identify emerging possibilities
3	**Risk Recognition** — Anticipate potential challenges
4	**Pattern Recognition** — Identify insights from early signals
5	**Timeline Calibration** — Develop realistic expectations

The most effective leaders combine ambitious vision with pragmatic understanding of technological realities, avoiding both over-optimism about near-term possibilities and under-estimation of longer-term implications.

Cross-Domain Integration

Digital workforce initiatives span traditional organizational boundaries, requiring leaders who can integrate perspectives and considerations across multiple domains:

- Connecting technological possibilities with business strategy and operational realities
- Integrating risk management, ethics, and operational considerations
- Balancing short-term implementation with long-term transformation
- Combining internal organizational perspectives with external stakeholder considerations
- Navigating both rational or analytical and emotional or cultural dimensions of change

Leaders who excel at this integration typically create more coherent implementation approaches and achieve better alignment across stakeholder groups than those operating from single functional perspectives.

Balanced Trust and Oversight

Effective leadership in the digital workforce era requires sophisticated calibration of trust and oversight, knowing when to empower digital systems and when to maintain human involvement. This capability includes:

- Developing nuanced understanding of AI capabilities and limitations in different contexts

- Creating appropriate governance frameworks that provide necessary oversight without inhibiting value

- Establishing clear accountability for decisions and outcomes in human-digital systems

- Building organizational confidence in digital capabilities while maintaining appropriate vigilance

- Evolving oversight approaches as capabilities mature and trust develops

Leaders who develop this balanced approach typically achieve better risk management and more efficient operations than those defaulting to either excessive caution or unwarranted confidence in digital systems.

Learning Leadership

Successful leaders must exemplify and foster continuous learning throughout their organizations to keep up with the evolving nature of digital workers:

- Demonstrating personal learning orientation and adaptation willingness

- Creating psychological safety for experimentation and candid feedback

- Establishing systematic knowledge-sharing mechanisms across initiatives
- Building organizational capabilities to absorb and apply emerging insights
- Balancing learning investment with performance expectations

Leaders who embody this learning orientation typically navigate technological evolution more successfully, respond to challenges more effectively, and sustain implementation momentum better than those approaching digital workforce initiatives with fixed mindsets or rigid planning approaches.

Human Development Focus

Perhaps most importantly, effective leaders maintain relentless focus on human development alongside technological advancement, recognizing that organizational success depends on evolving human capabilities in parallel with digital implementations:

- Investing in reskilling and development for changing roles
- Creating compelling career paths in evolving organizational structures
- Building cultural norms that support effective human-digital collaboration
- Recognizing and rewarding distinctly human contributions
- Maintaining meaningful work and purpose amid technological change

Leaders who prioritize this human dimension typically achieve greater stakeholder support, more successful adoption, and more sustainable organizational health than those focusing exclusively on technological implementation and efficiency gains.

Preparing for an Accelerating Future

As we conclude this exploration of digital workforce implementation, it is important to recognize that we stand at the beginning of a transformation that will likely accelerate and expand in the coming years. Several emerging developments warrant consideration in current planning and strategy.

Expanding Capability Horizon

While current agentic AI capabilities are already creating significant value, the trajectory of advancement suggests several important evolutions on the horizon:

- **Enhanced reasoning**: Increasingly sophisticated logical reasoning, causal understanding, and problem-solving capabilities

- **Multimodal integration**: More seamless understanding and generation across text, images, data, and potentially audio and video

- **Domain-specific depth**: Greater specialization in professional domains requiring deep expertise

- **Physical world connection**: Expanding integration with sensors, robotics, and physical systems

- **Self-improvement capabilities**: More autonomous learning, adaptation, and performance enhancement

Future of Agentic AI

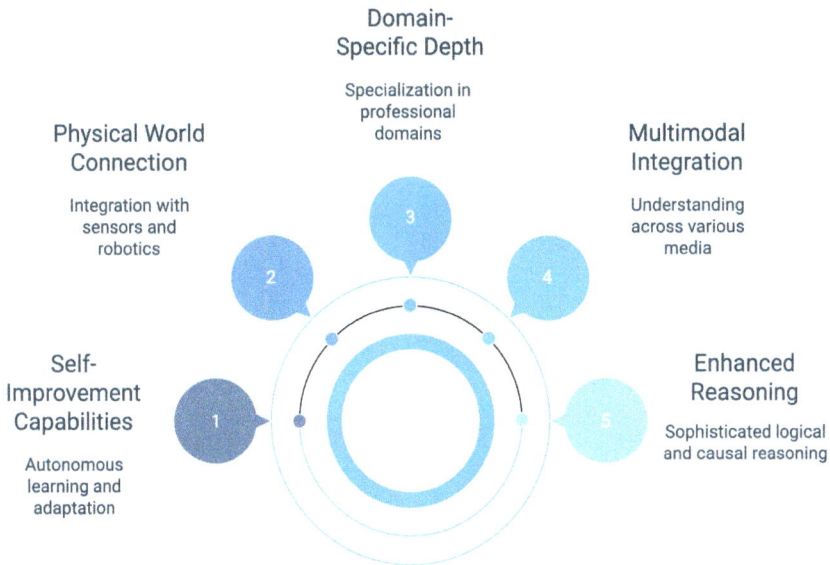

Domain-Specific Depth
Specialization in professional domains

Physical World Connection
Integration with sensors and robotics

Multimodal Integration
Understanding across various media

Self-Improvement Capabilities
Autonomous learning and adaptation

Enhanced Reasoning
Sophisticated logical and causal reasoning

Organizations should develop technical architectures, implementation approaches, and governance frameworks that can accommodate these evolving capabilities rather than optimizing exclusively for current technological realities.

Ecosystem Evolution

The digital workforce landscape will increasingly function as an interconnected ecosystem rather than a collection of isolated implementations. This evolution will include:

- **Platform consolidation**: Emerging dominant platforms providing comprehensive capabilities

- **Specialized solution proliferation**: Expanding array of domain and function-specific implementations

- **Standards and protocol development**: Evolution of integration and interoperability frameworks

- **Marketplace dynamics**: Emerging exchanges for capabilities, data, and specialized expertise
- **Regulatory framework expansion**: Development of governance standards and compliance requirements

Ecosystem Evolution

Platform consolidation	Dominant platforms with comprehensive capabilities are emerging.
Domain and function-specific implementations are expanding.	Specialized solutions
Standards development	Integration and interoperability frameworks are evolving.
Exchanges for capabilities, data, and expertise are emerging.	Marketplace dynamics
Regulatory expansion	Governance standards and compliance requirements are developing.

Organizations should position themselves strategically within this evolving ecosystem, developing partnership strategies, integration approaches, and positioning perspectives that extend beyond internal implementation considerations.

Workforce Transformation Acceleration

The impact on human roles, skills, and work experiences will likely accelerate and expand as digital workers become more capable and widespread. This transformation will include:

- **Role evolution acceleration**: Faster changes in job content and required capabilities

- **New role emergence**: Creation of entirely new positions focused on human-digital collaboration

- **Skill requirement shifts**: Growing emphasis on distinctly human capabilities like creativity, empathy, and judgment

- **Education system impacts**: Pressure for changes in learning and development approaches

- **Employment relationship evolution**: Potential shifts in work arrangements, career progression, and organizational structures

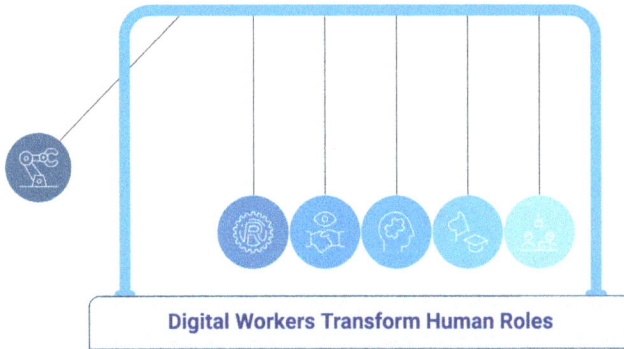

Digital Workers Transform Human Roles

Digital Workers	Role Evolution	New Roles	Skill Shifts	Education Impacts	Employment Evolution
Capable and widespread adoption	Faster changes in job content	Focus on human-digital collaboration	Emphasis on human capabilities	Changes in learning approaches	Shifts in work arrangements

Organizations should develop increasingly sophisticated approaches to workforce planning, development, and transition management that anticipate these accelerating changes rather than responding reactively as impacts emerge.

Strategic Differentiation Evolution

As digital workforce capabilities become more widely available, the basis of competitive differentiation will evolve beyond simple implementation to more sophisticated dimensions:

- **Integration sophistication**: How effectively organizations combine human and digital capabilities
- **Data and knowledge advantage**: Proprietary information that enhances digital worker performance
- **Implementation excellence**: Superior execution in design, deployment, and optimization
- **Cultural adaptability**: Organizational norms and behaviors that maximize human-digital collaboration
- **Strategic coherence**: Alignment of digital workforce initiatives with broader business direction

Strategic Differentiation Pyramid

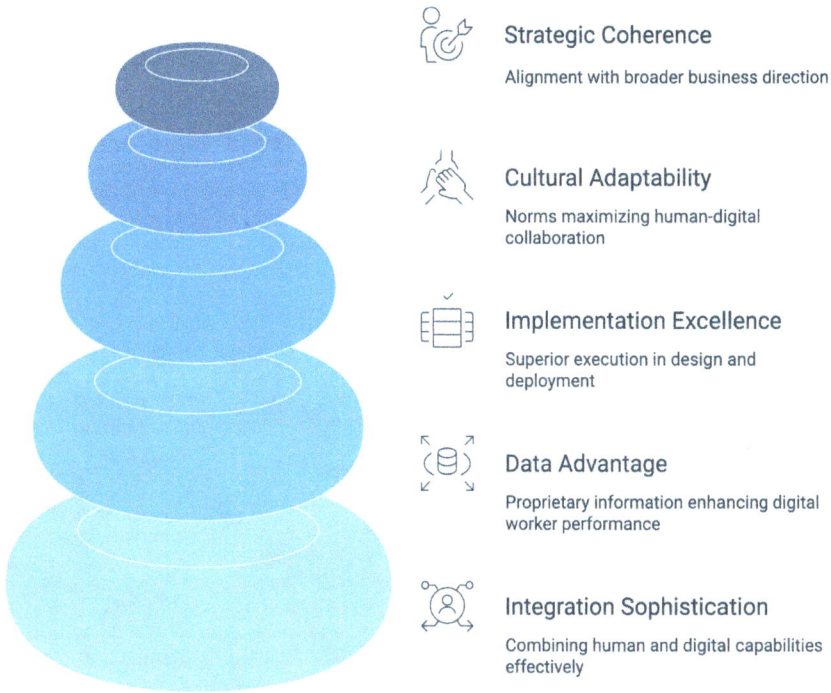

Strategic Coherence
Alignment with broader business direction

Cultural Adaptability
Norms maximizing human-digital collaboration

Implementation Excellence
Superior execution in design and deployment

Data Advantage
Proprietary information enhancing digital worker performance

Integration Sophistication
Combining human and digital capabilities effectively

Organizations should focus increasingly on these sustainable sources of advantage rather than relying on technological implementation alone.

Ethical and Societal Consideration Expansion

The broader implications of digital workforces for individuals, communities, and societies will receive growing attention, creating both constraints and opportunities:

- **Regulatory framework development**: Evolution of laws and standards governing AI implementation
- **Ethical expectation clarification**: Emerging consensus on appropriate use boundaries and principles

- **Distributional impact management**: Greater focus on who benefits from productivity enhancements

- **Environmental consideration integration**: Attention to energy, resource, and climate implications

- **Purpose and meaning evolution**: Broader societal conversation about human work in the digital age

Navigating the Ethical Landscape of Digital Workforces

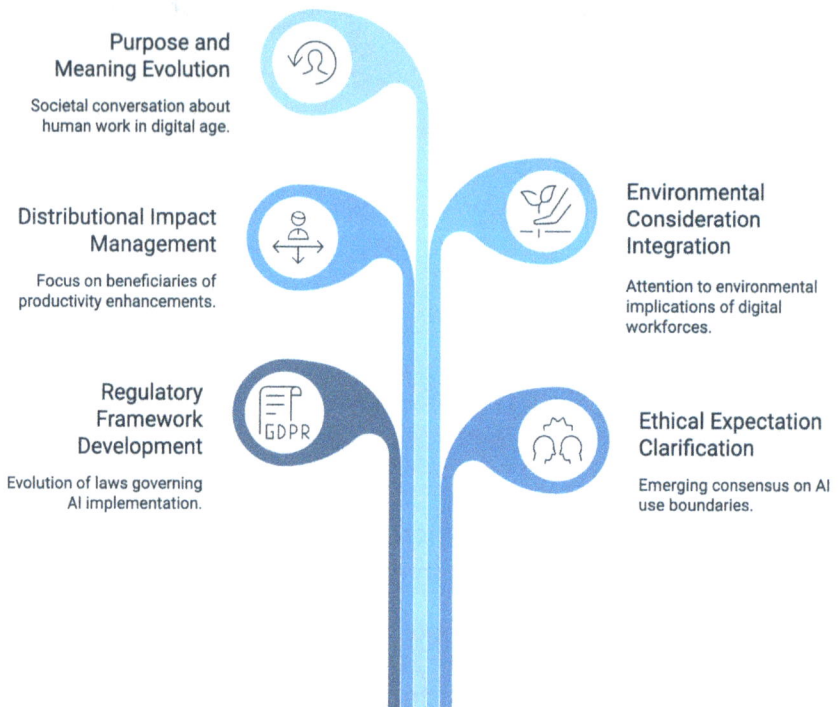

Purpose and Meaning Evolution
Societal conversation about human work in digital age.

Distributional Impact Management
Focus on beneficiaries of productivity enhancements.

Regulatory Framework Development
Evolution of laws governing AI implementation.

Environmental Consideration Integration
Attention to environmental implications of digital workforces.

Ethical Expectation Clarification
Emerging consensus on AI use boundaries.

Organizations should engage proactively with these considerations, viewing them as integral to implementation strategy rather than external constraints or secondary factors.

Final Thoughts: Embracing the Transformative Potential

As we conclude this exploration of digital workforce implementation, several final reflections can guide organizations navigating this transformation:

This is a marathon, not a sprint: While immediate implementation progress matters, sustainable success requires long-term perspective, systematic capability building, and strategic patience. Organizations should balance quick wins with development for enduring advantage.

Value creates permission: In a landscape of uncertainty, demonstrated value creation establishes organizational permission for continued investment and exploration. Practical results that benefit customers, employees, and stakeholders build support for broader transformation.

People remain central: Despite profound technological change, organizational success ultimately depends on human creativity, judgment, leadership, and relationships. The most successful digital workforce implementations enhance rather than diminish the human dimension.

Integration trumps implementation: The greatest value emerges not from isolated digital capabilities but from their integration into coherent business systems, organizational processes, and human workflows. Excellence in integration differentiates leaders from followers in this transformation.

Purpose provides direction: Amid rapid technological evolution, clarity about organizational purpose, values, and priorities provides essential direction for implementation decisions. Organizations should connect digital workforce initiatives to their core mission and identity.

The digital workforce era offers unprecedented opportunities for the enhancement of organizational capabilities, customer experiences, employee contributions, and competitive position. Organizations that approach this transformation thoughtfully, balancing technological possibility with human centrality, short-term progress with long-term vision, and efficiency gains with value creation, will define the next generation of organizational excellence.

We hope this book serves as a valuable companion on your journey toward building a digital workforce that creates sustainable value for all stakeholders and contributes positively to broader societal progress.

About the Author

Michael Fauscette is the Founder, CEO, and Chief Analyst at Arion Research, a globally recognized advisor and pioneering thought leader in digital transformation, agentic AI, and enterprise technology. With decades of experience as an industry analyst, executive, and board leader, Michael is a trusted voice for organizations navigating the future of work and intelligent automation. He previously served as Chief Research Officer at G2 and led IDC's global enterprise application software research. Michael shares his insights as a podcast host, industry speaker, and advisor to innovative technology companies.

Connect with Michael:
Website: yourdigitalworkforce.com
LinkedIn: linkedin.com/in/mfauscette
Podcast: disambiguationpod.com

www.ingramcontent.com/pod-product-compliance
Lightning Source LLC
Chambersburg PA
CBHW071537210326
41597CB00019B/3031